Lecture Notes in Computer Science 15171

Founding Editors

Gerhard Goos
Juris Hartmanis

Editorial Board Members

Elisa Bertino, *Purdue University, West Lafayette, IN, USA*
Wen Gao, *Peking University, Beijing, China*
Bernhard Steffen, *TU Dortmund University, Dortmund, Germany*
Moti Yung, *Columbia University, New York, NY, USA*

The series Lecture Notes in Computer Science (LNCS), including its subseries Lecture Notes in Artificial Intelligence (LNAI) and Lecture Notes in Bioinformatics (LNBI), has established itself as a medium for the publication of new developments in computer science and information technology research, teaching, and education.

LNCS enjoys close cooperation with the computer science R & D community, the series counts many renowned academics among its volume editors and paper authors, and collaborates with prestigious societies. Its mission is to serve this international community by providing an invaluable service, mainly focused on the publication of conference and workshop proceedings and postproceedings. LNCS commenced publication in 1973.

Maxime Chamberland · Tom Hendriks ·
Muge Karaman · Remika Mito · Nancy Newlin ·
S. Shailja · Elinor Thompson
Editors

Computational Diffusion MRI

15th International Workshop, CDMRI 2024
Held in Conjunction with MICCAI 2024
Marrakesh, Morocco, October 6, 2024
Proceedings

Editors
Maxime Chamberland
Eindhoven University of Technology
Eindhoven, The Netherlands

Tom Hendriks
Eindhoven University of Technology
Eindhoven, The Netherlands

Muge Karaman
University of Illinois Chicago
Chicago, IL, USA

Remika Mito
The University of Melbourne
Parkville, VIC, Australia

Nancy Newlin
Vanderbilt University
Nashville, TN, USA

S. Shailja
Stanford University
Palo Alto, CA, USA

Elinor Thompson
University College London
London, UK

ISSN 0302-9743 ISSN 1611-3349 (electronic)
Lecture Notes in Computer Science
ISBN 978-3-031-86919-8 ISBN 978-3-031-86920-4 (eBook)
https://doi.org/10.1007/978-3-031-86920-4

© The Editor(s) (if applicable) and The Author(s), under exclusive license to Springer Nature Switzerland AG 2025

This work is subject to copyright. All rights are solely and exclusively licensed by the Publisher, whether the whole or part of the material is concerned, specifically the rights of translation, reprinting, reuse of illustrations, recitation, broadcasting, reproduction on microfilms or in any other physical way, and transmission or information storage and retrieval, electronic adaptation, computer software, or by similar or dissimilar methodology now known or hereafter developed.
The use of general descriptive names, registered names, trademarks, service marks, etc. in this publication does not imply, even in the absence of a specific statement, that such names are exempt from the relevant protective laws and regulations and therefore free for general use.
The publisher, the authors and the editors are safe to assume that the advice and information in this book are believed to be true and accurate at the date of publication. Neither the publisher nor the authors or the editors give a warranty, expressed or implied, with respect to the material contained herein or for any errors or omissions that may have been made. The publisher remains neutral with regard to jurisdictional claims in published maps and institutional affiliations.

This Springer imprint is published by the registered company Springer Nature Switzerland AG
The registered company address is: Gewerbestrasse 11, 6330 Cham, Switzerland

If disposing of this product, please recycle the paper.

Preface

We are delighted to introduce the proceedings of the 2024 Computational Diffusion MRI (CDMRI) workshop. This workshop has been running for over a decade as a satellite event of the Medical Image Computing and Computer Assisted Intervention (MICCAI) conference. This year's CDMRI workshop continued as a platform to present some of the latest advancements in the field of diffusion MRI.

Diffusion-weighted imaging or diffusion MRI is a powerful approach that enables us to non-invasively probe microstructure of biological tissue. The past four decades have seen tremendous developments in the acquisition, analysis and application of diffusion MRI, delivering the promise of a clinical tool that can identify changes to tissue microstructure in the brain and beyond. Diffusion MRI also remains the only tool to be able to study the brain's white matter architecture in vivo, with developments in brain tractography demonstrating constant improvement. Despite these promising advances, there remain key challenges in the field of diffusion MRI. These challenges pertain to the clinical translatability of the technique, in difficulties with acquisition and accessibility, in expanding developments for diffusion MRI beyond the brain, as well as in tractography, where methodological variability still poses a huge hurdle for clinical translation. The submissions to CDMRI 2024 showcased novel solutions to the leading challenges that face the diffusion MRI community.

This year's CDMRI Workshop comprised three key themes centered around fantastic keynote lectures delivered by our invited speakers. Our first keynote talk, delivered by Godwin Ogbole, was fitting for our beautiful setting in Marrakesh, speaking of the challenges associated with bringing MRI research to Africa. The first theme—achieving more with less—was focused on papers that sought to make the best use of diffusion MRI data, using deep learning to improve quality and acquisition throughput, and to develop accurate models using the data. Our second keynote lecture was delivered by Sila Kurugol, and focused on challenges for diffusion MRI outside the brain. Papers falling within the second theme were similarly focused on diffusion MRI applications outside the brain, or perhaps outside the box, for example, using more unconventional modelling approaches for diffusion MRI. Finally, our third keynote speaker, Luc Florack, delivered a deep dive into geodesic tractography, kicking off the third theme on novel advances in tractography. Our final block of papers were all focused on novel developments in this area.

This workshop would not have been possible without the dedication of the Program Committee (listed below), who, through a double-blind peer review process, ensured the highest standard of publications. Of the 22 submissions we received, 19 were accepted after revisions, and 3 were rejected; each submission was reviewed by at least two members of the Program Committee. We extend our gratitude to everyone involved in the Program Committee. Finally, we wish to thank our Keynote Speakers (listed below), who delivered a series of truly enlightening lectures.

October 2024

Maxime Chamberland
Tom Hendriks
Muge Karaman
Remika Mito
Nancy Newlin
S. Shailja
Elinor Thompson

Organization

Organisers

Maxime Chamberland	Eindhoven University of Technology, The Netherlands
Tom Hendriks	Eindhoven University of Technology, The Netherlands
Muge Karaman	University of Illinois Chicago, USA
Remika Mito	University of Melbourne, Australia
Nancy Newlin	Vanderbilt University, USA
S. Shailja	Stanford University, USA
Elinor Thompson	University College London, UK

Program Committee

Abhishek Tiwari	Bennett University, India
Alberto De Luca	UMC Utrecht, The Netherlands
Alonso Ramirez-Manzanares	CIMAT A.C., Mexico
Andrey Zhylka	UMC Utrecht, The Netherlands
Antoine Theberge	Université de Sherbrooke, Canada
Bramsh Q Chandio	University of Southern California, USA
Chenyu Gao	Vanderbilt University, USA
Christophe Lenglet	University of Minnesota, USA
Daniel C. Moyer	Vanderbilt University, USA
Francois Rheault	Université de Sherbrooke, Canada
Graham Little	Université de Sherbrooke, Canada
Jon Haitz Legarreta Gorroño	Brigham and Women's Hospital, Mass General Brigham/Harvard Medical School, USA
Jose Pedro Manzano	Patron University of Nottingham, UK
Kurt G. Schilling	Vanderbilt University, USA
Lars Smolders	Elizabeth-TweeSteden Ziekenhuis, The Netherlands
Lianrui Zuo	Vanderbilt University, USA
Marco Palombo	Cardiff University, UK
Marco Pizzolato	DTU, Denmark
Marta M. Correia	University of Cambridge, UK
Nagesh Adluru	University of Wisconsin, Madison, USA

Nazirah Mohd Khairi   Vanderbilt University, USA
Neil Oxtoby   University College London, UK
Pamela Guevara   Universidad de Concepción, Chile
Pew-Thian Yap   UNC Chapel Hill, USA
Robert E. Smith   Florey Institute of Neuroscience and Mental Health, Australia
Süheyla Cetin-Karayumak   Harvard Medical School and Brigham and Women's Hospital, USA
Suyash P. Awate   Indian Institute of Technology Bombay, India
Thomas Schultz   University of Bonn, Germany
Yaël Balbastre   University College London, UK
Ye Wu   Nanjing University of Science and Technology, China
Zhiyuan Li   Vanderbilt University, USA

Keynote Speakers

Godwin Ogbole   University of Ibadan, Nigeria
Sila Kurugol   Harvard Medical School, USA
Luc Florack   Eindhoven University of Technology, The Netherlands

Contents

Super-Resolution of Diffusion-Weighted Images via TDI-Conditioned
Diffusion Model 1
   *Jiquan Ma, Yujun Teng, Geng Chen, Haotian Jiang, Kai Zhang,
Feihong Liu, Islem Rekik, and Dinggang Shen*

Diffusion-Based Gray-White Matter Mapping for Quantitative
Tractography in Glioma Patients 12
   *Lars Smolders, Maxime Chamberland, Geert-Jan Rutten,
Wouter De Baene, Remco van der Hofstad, and Luc Florack*

Ground-Truth Effects in Learning-Based Fiber Orientation Distribution
Estimation in Neonatal Brains 24
   *Rizhong Lin, Hamza Kebiri, Ali Gholipour, Yufei Chen,
Jean-Philippe Thiran, Davood Karimi, and Meritxell Bach Cuadra*

Synthesizing 3D Axon Morphology: Springs are All We Need 35
   Ruiqi Cui, J. Andreas Bærentzen, and Tim B. Dyrby

Randomly COMMITting: Iterative Convex Optimization
for Microstructure-Informed Tractography 47
   Sanna Persson, Xinyi Wan, and Rodrigo Moreno

AID-DTI: Accelerating High-Fidelity Diffusion Tensor Imaging
with Detail-Preserving Model-Based Deep Learning 60
   *Wenxin Fan, Jian Cheng, Cheng Li, Jing Yang, Ruoyou Wu, Juan Zou,
and Shanshan Wang*

Multi-dimensional Parameter Space Exploration for Streamline-Specific
Tractography 72
   Ruben Vink, Anna Vilanova, and Maxime Chamberland

Cross-Domain Fiber Cluster Shape Analysis for Language Performance
Cognitive Score Prediction 84
   *Yui Lo, Yuqian Chen, Dongnan Liu, Wan Liu, Leo Zekelman,
Fan Zhang, Yogesh Rathi, Nikos Makris, Alexandra J. Golby,
Weidong Cai, and Lauren J. O'Donnell*

Can Transfer Learning Improve Supervised Segmentation of White Matter
Bundles in Glioma Patients? 95
 Chiara Riccardi, Sofia Ghezzi, Gabriele Amorosino, Luca Zigiotto,
 Silvio Sarubbo, Jorge Jovicich, and Paolo Avesani

Image Quality Transfer of Diffusion MRI Guided By High-Resolution
Structural MRI 106
 Alp G. Cicimen, Henry F. J. Tregidgo, Matteo Figini,
 Eirini Messaritaki, Carolyn B. McNabb, Marco Palombo,
 C. John Evans, Mara Cercignani, Derek K. Jones,
 and Daniel C. Alexander

QID^2: An Image-Conditioned Diffusion Model for Q-Space Up-Sampling
of DWI Data 119
 Zijian Chen, Jueqi Wang, and Archana Venkataraman

Ts-FWE: Token-Aware Single-Shell Free Water Estimation for Brain
Diffusion MRI 132
 Tianyuan Yao, Derek Archer, Zhiyuan Li, Leon Y. Cai,
 Praitayini Kanakaraj, Nancy Newlin, Quan Liu, Ruining Deng,
 Can Cui, Shunxing Bao, Kurt Schilling, Bennett A. Landman,
 and Yuankai Huo

Assessing Early Motor System Degeneration in the Spinal Cord of ALS
Patients Using Diffusion MRI: An Exploratory Study 143
 Alexandra Ford, Andrew W. Barritt, and Samira Bouyagoub

RobNODDI: Robust NODDI Parameter Estimation with Adaptive
Sampling Under Continuous Representation 153
 Taohui Xiao, Jian Cheng, Wenxin Fan, Jing Yang, Cheng Li,
 Enqing Dong, and Shanshan Wang

Introducing QuantConn: Overcoming Challenging Diffusion Acquisitions
with Harmonization 164
 Nancy Newlin, Kurt Schilling, Serge Koudoro, Bramsh Qamar Chandio,
 Praitayini Kanakaraj, Daniel Moyer, Claire E. Kelly,
 Sila Genc, Joseph Yuan-Mou Yang, Ye Wu, Nagesh Adluru,
 Vishwesh Nath, Sudhir Pathak, Walter Schneider, Anurag Gade,
 William Consagra, Yogesh Rathi, Tom Hendriks, Anna Vilanova,
 Maxime Chamberland, Tomasz Pieciak, Dominika Ciupek,
 Antonio Tristán Vega, Santiago Aja-Fernández, Maciej Malawski,
 Gani Ouedraogo, Julia Machnio, Paul M. Thompson, Neda Jahanshad,
 Eleftherios Garyfallidis, and Bennett Landman

Learning Low-Rank Tensor Approximation for GPU-Based Tractography 175
   Johannes Gruen and Thomas Schultz

Deep Multivariate Autoencoder for Capturing Complexity in Brain
Structure and Behaviour Relationships 185
   Gabriela Gómez Jiménez and Demian Wassermann

Heritability and Genetic Correlations Along the Corticospinal Tract 197
   Iyad Ba Gari, Ravi R. Bhatt, Fang-Chang Yeh, and Neda Jahanshad

Corpus Callosum Parcellation Methods: What Can Tractography Tell Us
About Them? 210
   *Caio Santana, Claudio Román, Simone Appenzeller, Pamela Guevara,
   and Leticia Rittner*

Author Index 223

Super-Resolution of Diffusion-Weighted Images via TDI-Conditioned Diffusion Model

Jiquan Ma[1], Yujun Teng[1], Geng Chen[2(✉)], Haotian Jiang[2], Kai Zhang[3], Feihong Liu[3,4], Islem Rekik[5], and Dinggang Shen[3]

[1] School of Computer Science and Technology, Heilongjiang University, Harbin, China
[2] National Engineering Laboratory for Integrated Aero-Space-Ground-Ocean Big Data Application Technology, School of Computer Science and Engineering, Northwestern Polytechnical University, Xi'an, China
geng.chen.cs@gmail.com
[3] School of Biomedical Engineering and State Key Laboratory of Advanced Medical Materials and Devices, ShanghaiTech University, Shanghai, China
[4] School of Information Science and Technology, Northwest University, Xi'an, China
[5] BASIRA Lab, Imperial-X and Department of Computing, Imperial College London, London, UK

Abstract. Diffusion-Weighted Imaging (DWI) is a significant technique for studying white matter. However, it suffers from low-resolution obstacles in clinical settings. Post-acquisition Super-Resolution (SR) can enhance the resolution of DWIs and has gained increasing research interest in recent years. An advanced generative model, the Diffusion Model (DM), exhibits particularly promising performance in image SR. However, effective conditions are required to bootstrap the DM for DWI SR. To this end, we proposed the first DM-based DWI SR model with two effective conditions based on low-solution DWIs and Track Density Imaging (TDI) maps, which possess rich high-resolution prior knowledge Additionally, we consider another condition based on features from low-resolution DWIs. These two conditions are integrated into our model, which comprises three components: DWI Resolution Enhancer (DRE), DWI Feature Extractor (DFE), and TDI Feature Extractor (TFE). DRE combines low-resolution DWI features from DFE with TDI features from TFE to progressively generate high-resolution DWIs. We performed extensive experiments on DWIs of normal subjects from human connectome projects and patients with Parkinson's disease. The results demonstrate that our model outperforms existing DWI SR models, both qualitatively and quantitatively.

Keywords: Diffusion-Weighted Imaging · Super-Resolution · Conditional Diffusion Model · Track Density Imaging

---

This work was supported in part by the National Natural Science Foundation Project (62201465) and the Heilongjiang Provincial Natural Science Foundation Project (LH2021F046).

1 Introduction

Diffusion-Weighted Imaging (DWI) is a non-invasive technique that probes the motion of water molecules in soft tissues, providing valuable information for studying brain white matter [11]. However, due to the limited acquisition time, DWI suffers from low-resolution issues. Meanwhile, low resolution can lead to severe partial volume effects, which cause detrimental effects on subsequent clinical analyses [10]. To address the challenges posed by low resolution in DWIs, researchers have delved into the DWI Super-Resolution (SR) techniques, which aim to enhance image quality and offer more structural details [7,10,26]. The primary objective of SR is to reconstruct high-resolution DWIs from low-resolution DWIs by harnessing either model-driven or data-driven techniques.

The conventional SR methods of DWIs are based on interpolation techniques [5,6,13] and non-local means regularization [4,22]. Compared with these methods, deep learning techniques have gained significant attention in the field of SR [16,17,30], recently. He et al. demonstrated the viability of Convolutional Neural Networks (CNN) for image SR and introduced the Super Resolution Convolutional Neural Network (SRCNN) [9]. Elasid et al. extended this concept by applying the two-dimensional SRCNN to enhance the resolution of DWIs [10]. However, this approach employs a shallow CNN and only low-resolution DWIs for SR. Tian et al. proposed a modified DWI SR network, SR for diffusion tensor MRI, which integrates residual learning and multi-contrast imaging into SRCNN [27]. Tanno et al. expanded the scope by extending the original two-dimensional efficient sub-pixel convolutional neural network to perform SR on three-dimensional DWIs, incorporating uncertainty into the learning process [26]. In a different vein, Albay et al. introduced a data-driven SR generative adversarial network to enhance the spatial resolution of DWIs [1]. More recently, Muhammad et al. combined GoogleLeNet and ResNet to propose an Inception-ResNet network, IRMIRS, specifically designed for DWI SR [23]. Despite the progress of existing methods, as an ill-posed problem, accurate resolution enhancement of DWIs is still a challenging task.

Diffusion Model (DM) is a type of generative deep learning model that has gained significant attention in recent years and has shown promising performance in the SR of natural images [15]. Unlike other generative models, the DM employs a probabilistic modeling approach based on diffusion processes and achieves promising performance in both realistic generation and model convergence. However, the DM has not been utilized for the SR of DWIs. Designing a DM-based DWI SR method possesses several challenges, where a major one is how to design effective conditions for the DM for high-quality enhancement of the DWI resolution. A promising choice for the condition of DM is the Track Density Imaging (TDI) [3], which is derived from low-resolution DWIs and contains rich high-resolution information. Specifically, TDIs are obtained by a post-processing method that utilizes tractography to reveal structures beyond the acquired imaging voxel resolution and can recover microstructure information lost due to low resolution [3].

To this end, we propose the first DM-based SR model for DWIs, called the TDI-Conditioned Diffusion Model (TCDM). Our TCDM possesses three components, including the DWI Resolution Enhancer (DRE), DWI Feature Extractor (DFE), and TDI Feature Extractor (TFE). The DRE integrates an attention component into the basic unit of ResUnet [8] to construct the Attention Residual Block (ARB). The DFE and TFE extract the features of low-resolution DWIs and TDIs to incorporate these two conditions into TCDM. The technical contributions of our work are summarized as follows:

- We are the first to propose a DM-based DWI SR model, which incorporates TDIs and low-resolution DWIs in the DM are conditions for enhancing the resolution of DWIs.
- We design (i) TFE to extract the features from TDIs containing rich high-resolution information and (ii) DFE to extract the features from low-resolution DWIs containing essential structural information.
- We design an effective network backbone, DRE, which consists of multiple ARBs for gradually recovering high-resolution DWIs.

Through extensive experimental validation with data from the Human Connectome Project (HCP) [28] and a Parkinson's Disease (PD) [2] study, the proposed TCDM outperforms the most advanced DWI SR methods across different datasets, both qualitatively and quantitatively. Our code is publicly available at GitHub[1].

2 Method

2.1 TDI-Conditioned Diffusion Model

Optimization of Our Model. We optimize the proposed SR module $\text{TCDM}_\theta(\cdot)$ that takes two conditional images, a low-resolution DWI x_d and the corresponding TDI x_t, as input along with a noisy target image $\tilde{y}$, which is a high-resolution image that has been diffused through several steps. To prevent parameter fixing, the number of diffusion steps is randomly sampled from the total number of diffusion steps. Our objective function for training $\text{TCDM}_\theta(\cdot)$ is defined as:

$$E\|\text{TCDM}_\theta(x_d, x_t, \underbrace{\sqrt{\gamma}y_0 + \sqrt{1-\gamma}\epsilon}_{\tilde{y}}, \gamma) - \epsilon\|^2, \qquad (1)$$

where γ is the variance of Gaussian distribution [25], ϵ denotes the standard Gaussian distribution, $E\|\cdot\|^2$ represents the calculated mean square error, and θ refers to the weights of TCDM. The goal of the model is to predict the Gaussian distribution noise added at the $t-th$ step.

The training phase of TCDM is depicted in Algorithm 1, where $p(x_d, x_t, y)$ represents a minibatch of samples randomly drawn from the training set, $p(t)$ represents the cumulative multiplicative variance over the first t diffusion steps of the diffusion process, and $N(0, 1)$ denotes the standard normal distribution.

---
[1] https://github.com/yjtengAlex/TCDM_master.

Algorithm 1. Training of $\text{TCDM}_\theta(\cdot)$

Input: Low-resolution DWI x_d, corresponding TDI x_t, high-resolution DWI y_0, and diffusion steps t.
Output: Trained network parameters
1: **while** true **do**
2:    $(x_d, x_t, y_0) \sim p(x_d, x_t, y)$
3:    $\gamma \sim p(t)$
4:    $\epsilon \sim N(0, 1)$
5:    Take a gradient descent step on
6:       $\nabla_\theta \, \|\text{TCDM}_\theta(x_d, x_t, \sqrt{\gamma}y_0 + \sqrt{1-\gamma}\epsilon, \gamma) - \epsilon\|^2$
7:    **if** converged **then**
8:       break
9:    **end if**
10: **end while**
11: **return** the parameters of model $\text{TCDM}_\theta(\cdot)$

Super-Resolution Process. We extend Denoising Diffusion Implicit Models (DDIM) [24] for performing SR and defining the back diffusion process as follows:

$$y_{t-1} = \sqrt{\gamma_{t-1}} \left(\frac{y_t - \sqrt{1-\gamma_t} \cdot \text{TCDM}_\theta(x_d, x_t, y_t, \gamma_t)}{\sqrt{\gamma_t}} \right) \\ + \sqrt{1 - \gamma_t - \sigma_t^2} \cdot \text{TCDM}_\theta(x_d, x_t, y_t, \gamma_t) + \sigma_t \epsilon. \quad (2)$$

where γ_t is the cumulative product of the variance of Gaussian distribution up to t steps [25], y_t is the high-resolution image diffused for t steps, and σ_t is a DDIM variable defined as follows:

$$\sigma_t = \sqrt{(1-\gamma_{t-1})/(1-\gamma_t)} \cdot \sqrt{1 - \gamma_t/\gamma_{t-1}}. \quad (3)$$

Based on Eq. (2), we can use the trained model to perform the SR processing of DWIs. Specifically, the model predicts the noise added by the diffusion at step t, which is then subtracted in a Bayesian fashion [15]. Next, we obtain a DWI that has been diffused for $t-1$ steps. By performing this back diffusion process step by step, the noisy image is restored to a noise-free high-resolution image.

2.2 Model Components

Figure 1(A) represents the forward process of DM. During this process, features are extracted from low-resolution DWIs and associated TDIs through DFE and TFE, respectively. Guided by these two conditions, noise is gradually added, ultimately leading to an approximate pure Gaussian distribution. It is important to note that the three components of TCDM are trained simultaneously.

DWI Feature Extractor. We designed a DWI Feature Extractor (DFE) to take full advantage of low-resolution DWIs, as shown in the lower left of

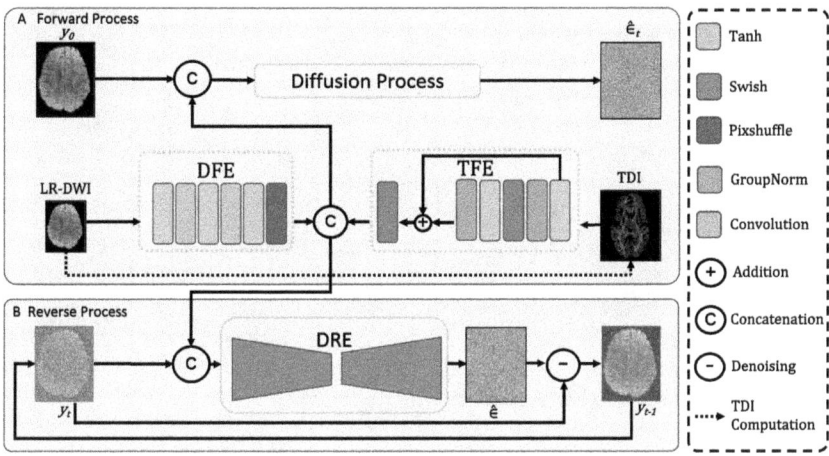

Fig. 1. An Overview of the TCDM. The TCDM consists of three modules: DFE, TFE, and DRE. (A) indicate forward diffusion process. (B) denotes a super-resolution back diffusion process.

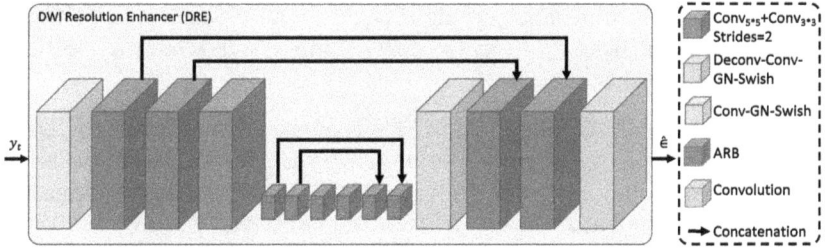

Fig. 2. Schematic diagram of the DRE.

Fig. 1(A). The feature extractor is mainly composed of several convolutional layers and pixel shuffle layers. Convolutional layers extract the features to learn the pattern and structure of the low-resolution DWI, and the pixel shuffle layers are used to map a low-resolution DWI into the high-resolution space. With such a feature extractor structure, we can effectively extract and utilize features in low-resolution DWIs to achieve higher-quality image reconstruction.

For DFE, we represent the size of the convolution kernel and the number of convolution kernels in the i-th layer as f_i and n_i, respectively. For the implementation details of DFE, we design DFE with $(f_1, n_1) = (5, 64), (f_2, n_2) = (3, 32), (f_3, n_3) = (3, 4)$, and Tanh activation function for all convolutional layers. These implementations result in a feature map of size $H \times W \times 4$. The feature map is then passed through the pixel shuffle layer to obtain a feature map of size $2H \times 2W \times 1$.

TDI Feature Extractor. TDIs contain rich high-resolution information of DWIs in the fiber tract domain [3]. It should be noted that, although TDIs

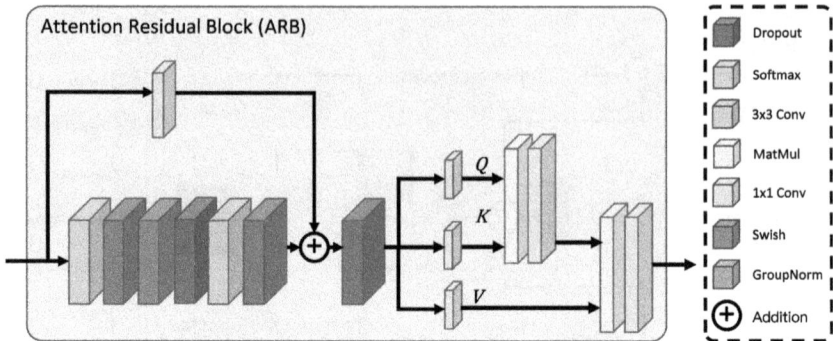

Fig. 3. Schematic diagram of the ARB.

have the same resolution as high-resolution DWIs, they are derived from low-resolution DWIs. We used MRtrix [3] to convert DWIs into TDIs. To make the SR images contain more structural information, we designed the TDI Feature Extractor (TFE) module to extract the features from TDIs as another condition of our TCDM, as shown in the lower right of Fig. 1(A). Specifically, TFE consists of two convolutional layers, $(f_1, n_1) = (3, 64)$ and $(f_2, n_2) = (3, 1)$, with Swish [18] activation functions.

DWI Resolution Enhancer. The DWI Resolution Enhancer (DRE) is designed as a ResUnet structure with one downsampling and multiple skip connections, as shown in Fig. 2. The inputs of DRE are the feature maps of low-resolution DWIs, the feature maps of TDIs, and the high-resolution DWIs with noise. These inputs are processed as follows: (i) They are first fused and then passed through the bottleneck layer to increase the number of channels. (ii) The resulting feature maps are then fed to two ARBs followed by a subsampling layer to reduce the size of the feature maps. (iii) Next, they go through six ARBs, where the last two ARBs have skip connections with the first two ARBs. (iv) Deconvolutional layers are then used to upscale the feature map dimensions. Next are two ARBs, each of which has a jump connection to the ARBs that preceded the downsampling operation. (v) The number of channels is adjusted through the tail layer and the Gaussian distributed noise is the output. (vi) The downsampling layer is borrowed from the YOLOv3 [31] and is achieved by two different sizes of convolution operations with a step size of two.

Attention Residual Block. As is shown in Fig. 3, Attention Residual Block (ARB) is an improved residual block that integrates a self-attention component with residual connections. This enhances the attention of the network to structural information. Due to the small batch size, the traditional batch normalization is no longer applicable, so we adopt a new normalization strategy, called group normalization [29]. Additionally, we use the Swish [18] activation function, which is effective in improving model performance, and add dropout layers to alleviate overfitting.

Table 1. Quantitative comparison in "PSNR/SSIM" format for 2× DWI SR. We compare the results from HCP data, PD data, and corresponding Fractional Anisotropy (FA) images. The best results are in **bold**.

Method	HCP-DWI		PD-DWI		HCP-FA		PD-FA	
	PSNR ↑	SSIM ↑	PSNR ↑	SSIM ↑	PSNR ↑	SSIM ↑	PSNR ↑	SSIM ↑
VDSR	27.57	0.9395	20.92	0.8729	46.60	0.9277	23.25	0.8025
EDSR	24.68	0.9244	21.33	0.8686	40.92	0.9483	22.38	0.8092
ESPCN	26.45	0.9436	22.24	0.8841	48.24	0.9863	24.57	0.8538
SRGAN	28.71	0.9610	22.35	0.8942	46.65	0.9829	23.87	0.8362
IRMIRS	25.10	0.9346	21.70	0.8786	41.04	0.9486	22.37	0.8043
TCDM	**31.03**	**0.9732**	**23.94**	**0.9232**	**48.65**	**0.9892**	**24.67**	**0.8556**

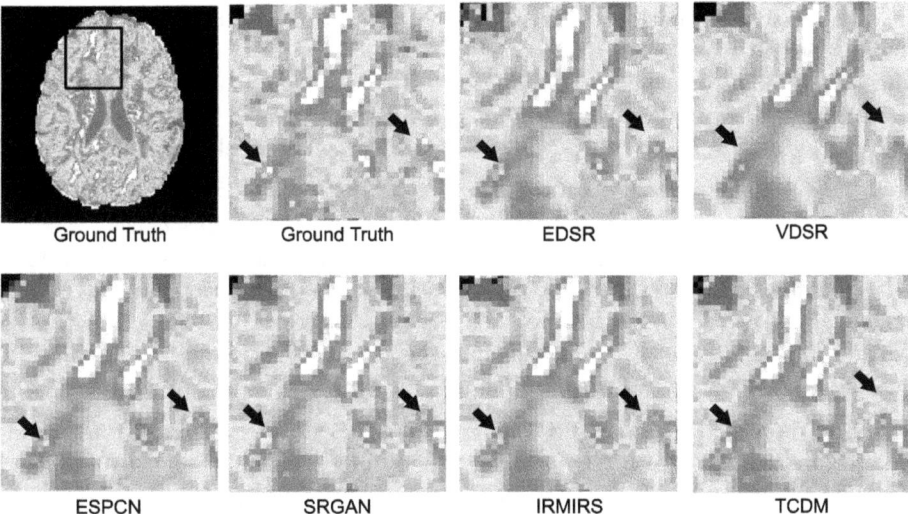

Fig. 4. Visual comparison of DWIs. Our model outperforms comparison models, especially for the regions marked by arrows.

3 Experiments

Dataset. In the experiment, we selected 15 subjects from the HCP [28] with ten subjects being used as the training set and five subjects as the testing set. To simulate a clinical scenario, we uniformly chose 30 gradient directions with a b-value of 1000 s/mm^2. We performed the brain extraction followed by noise reduction with Patch2Self [12]. Each DWI had a size of $145 \times 174 \times 145$, which we cropped to $112 \times 144 \times 112$ to remove the background. By averaging adjacent voxels in the DWI, we obtained a low-resolution DWI with a resolution of $56 \times 72 \times 56$. We used the average b_0 image for normalization.

Table 2. Quantitative results for ablation study. We compared the PSNR and SSIM of DWIs as well as the PSNR and SSIM of FA images. The best results are in **bold**.

Version	DFE	TFE	DWI-PSNR ↑	DWI-SSIM ↑	FA-PSNR ↑	FA-SSIM ↑
(A)			26.68	0.9287	43.49	0.9671
(B)	✓		27.52	0.9478	47.35	0.9843
(C)	✓	✓	**31.03**	**0.9732**	**48.65**	**0.9892**

The TCDM is a two-dimensional network, and we extracted two-dimensional slices from the sagittal, coronal, and axial directions. The training set contains 110,400 slices and the testing set contains 55,200 slices. During the inference, the results of three views are averaged to get the final SR result. Besides, we use the Peak Signal-to-Noise Ratio (PSNR) and Structural Similarity (SSIM) for quantitative evaluation. To validate the generalization ability of TCDM, we conduct testing using a PD dataset[2], which consists of DWIs of five PD subjects. The dimension and voxel size of the DWI are $112 \times 112 \times 50$ and two millimeters, respectively. The number of two-dimensional slices obtained in sagittal, coronal, and axial views is 41, 100. It is worth noting that our model is exclusively used for testing purposes only without training for the PD dataset.

Implementation Details. We use three RTX 2080 Ti graphics cards for training, each equipped with 11 GB of memory and running CUDA version 11.6. Additionally, we employ Python version 3.9.18 and PyTorch version 1.13.1. The optimizer chosen is the AdamW optimizer, with the weight decay for the initial learning rate set to $1e^{-4}$. To further improve the training process, we adhere to the strategy of learning rate warm-up [14].

Results. The proposed TCDM was compared with cutting-edge SR methods on the HCP dataset and PD dataset. The comparison methods include Very Deep Convolutional Networks (VDSR) [19], Enhanced Deep Residual Networks (EDSR) [21], Efficient Sub-Pixel Convolutional Neural Network (ESPCN) [26], Super-Resolution GAN (SRGAN) [1,20], and Inception-ResNet-Based Network for MRI Image Super-Resolution (IRMIRS) [23].

The results, as shown in Table 1, indicate that TCDM achieves the best performance. Compared with VDSR, EDSE, ESPCN, SRGAN and IRMIRS, our TCDM has a PSNR improvement of about 3.46, 6.35, 4.58, 2.32 and 5.93 dB on the HCP dataset with 2 times magnification factor, respectively. Furthermore, we show the visualization results in Fig. 4. As can be observed, our TCDM outperforms comparison models and provides DWIs that are the closest to the ground truth.

Ablation Study. We performed extensive ablation experiments to verify the effectiveness of DFE and TFE. Table 2 shows the results for DWIs and FA images of the five testing subjects. We constructed a basic version (A), which refers to the absence of DFE and TFE in the model.

---
[2] https://www.nitrc.org/projects/parktdi.

Effectiveness of DWI Feature Extractor: Version (B) has no TFE but with DFE. Compared with the baseline (A), although DWI-PSNR and DWI-SSIM increased by 0.84 dB and 0.0191, FA-PSNR and FA-SSIM have a tremendous boost increase by 3.86 dB, and 0.0172. Therefore, the designed DFE can effectively improve performance.

Effectiveness of TDI Feature Extractor: We verify the effectiveness of TFE by creating an ablated version (C) in which both DFE and TFE are components of the model. Compared with version (B), the DWI-PSNR and DWI-SSIM have remarkable improvement increased by 3.51 and 0.0254. Meanwhile, the FA-PSNR and FA-SSIM also increased by 1.30 and 0.0049. Consequently, the TFE can further enhance the performance.

4 Conclusion

In this work, we proposed a novel model, TCDM, which is the first diffusion model to enhance the resolution of DWIs. The TCDM consists of three key modules: (i) The DFE module extracts the key features of low-resolution DWIs to lay the foundation for SR reconstruction, (ii) The TFE module extracts the features from TDIs and helps to recover the structural details of DWIs, and (iii) The DRE module fuses the features of low-resolution DWIs and TDIs to gradually recover high-resolution DWIs. Extensive experiments demonstrate that our TCDM achieves remarkable results in DWI SR.

References

1. Albay, E., Demir, U., Unal, G.: Diffusion MRI spatial super-resolution using generative adversarial networks. In: PRIME, pp. 155–163. Springer, Cham (2018)
2. Bajaj, S., et al.: Diffusion-weighted MRI distinguishes Parkinson disease from the parkinsonian variant of multiple system atrophy: a systematic review and meta-analysis. PLoS ONE **12**(12), e0189897 (2017)
3. Calamante, F., Tournier, J.D., Jackson, G.D., Connelly, A.: Track-density imaging (TDI): super-resolution white matter imaging using whole-brain track-density mapping. Neuroimage **53**(4), 1233–1243 (2010)
4. Chen, G., Dong, B., Zhang, Y., Lin, W., Shen, D., Yap, P.T.: XQ-SR: joint x-q space super-resolution with application to infant diffusion MRI. Med. Image Anal. **57**, 44–55 (2019)
5. Chen, G., Dong, B., Zhang, Y., Shen, D., Yap, P.T.: q-Space upsampling using x-q space regularization. In: MICCAI, pp. 620–628 (2017)
6. Chen, G., Dong, B., Zhang, Y., Lin, W., Shen, D., Yap, P.-T.: Angular upsampling in infant diffusion MRI using neighborhood matching in x-q space. Front. Neuroinform. **12**, 57 (2018)
7. Coupé, P., Yger, P., Prima, S., Hellier, P., Kervrann, C., Barillot, C.: An optimized blockwise nonlocal means denoising filter for 3-D magnetic resonance images. IEEE Trans. Med. Imaging **27**(4), 425–441 (2008)

8. Diakogiannis, F.I., Waldner, F., Caccetta, P., Wu, C.: ResUNet-a: a deep learning framework for semantic segmentation of remotely sensed data. ISPRS J. Photogramm. Remote. Sens. **162**, 94–114 (2020)
9. Dong, C., Loy, C.C., He, K., Tang, X.: Learning a deep convolutional network for image super-resolution. In: ECCV, pp. 184–199. Springer, Cham (2014)
10. Elsaid, N.M., Wu, Y.C.: Super-resolution diffusion tensor imaging using SRCNN: a feasibility study. In: EMBC, pp. 2830–2834. IEEE (2019)
11. Essayed, W.I., Zhang, F., Unadkat, P., Cosgrove, G.R., Golby, A.J., O'Donnell, L.J.: White matter tractography for neurosurgical planning: a topography-based review of the current state of the art. NeuroImage: Clin. **15**, 659–672 (2017)
12. Fadnavis, S., Batson, J., Garyfallidis, E.: Patch2Self: denoising diffusion MRI with self-supervised learning. In: NeurIPS (2020)
13. Gulati, T., Sinha, H.: Interpreting low resolution MRI images using polynomial based interpolation. Int. J. Eng. Trends Technol. (IJETT) **10**, 626–631 (2014)
14. He, K., Zhang, X., Ren, S., Sun, J.: Deep residual learning for image recognition. In: CVPR, pp. 770–778 (2016)
15. Ho, J., Jain, A., Abbeel, P.: Denoising diffusion probabilistic models. In: NeurIPS, vol. 33, pp. 6840–6851 (2020)
16. Hong, Y., Chen, G., Yap, P.T., Shen, D.: Multifold acceleration of diffusion MRI via deep learning reconstruction from slice-undersampled data. In: IPMI, pp. 530–541 (2019)
17. Huang, S., et al.: Super-resolution reconstruction of fetal brain MRI with prior anatomical knowledge. In: IPMI (2023)
18. Kiaei, A.A., et al.: Active Identity Function as Activation Function (2023)
19. Kim, J., Lee, J.K., Lee, K.M.: Accurate image super-resolution using very deep convolutional networks. In: Proceedings of the IEEE Conference on Computer Vision and Pattern Recognition, pp. 1646–1654 (2016)
20. Ledig, C., et al.: Photo-realistic single image super-resolution using a generative adversarial network. In: CVPR, pp. 4681–4690 (2017)
21. Lim, B., Son, S., Kim, H., Nah, S., Mu Lee, K.: Enhanced deep residual networks for single image super-resolution. In: Proceedings of the IEEE Conference on Computer Vision and Pattern Recognition Workshops, pp. 136–144 (2017)
22. Manjón, J.V., Carbonell-Caballero, J., Lull, J.J., García-Martí, G., Martí-Bonmatí, L., Robles, M.: MRI denoising using non-local means. Med. Image Anal. **12**(4), 514–523 (2008)
23. Muhammad, W., Bhutto, Z., Masroor, S., Shaikh, M.H., Shah, J., Hussain, A.: IRMIRS: inception-ResNet-based network for MRI image super-resolution. CMES-Comput. Model. Eng. Sci. **136**(2) (2023)
24. Nichol, A.Q., Dhariwal, P.: Improved denoising diffusion probabilistic models. In: International Conference on Machine Learning, pp. 8162–8171. PMLR (2021)
25. Saharia, C., Ho, J., Chan, W., Salimans, T., Fleet, D.J., Norouzi, M.: Image super-resolution via iterative refinement. IEEE Trans. Pattern Anal. Mach. Intell. **45**(4), 4713–4726 (2022)
26. Tanno, R., et al.: Bayesian image quality transfer with CNNs: exploring uncertainty in dMRI super-resolution. In: MICCAI, pp. 611–619. Springer, Cham (2017)
27. Tian, Q., et al.: SRDTI: deep learning-based super-resolution for diffusion tensor MRI. arXiv preprint arXiv:2102.09069 (2021)
28. Van Essen, D.C., et al.: The WU-Minn human connectome project: an overview. Neuroimage **80**, 62–79 (2013)

29. Wu, Y., He, K.: Group normalization. In: ECCV, pp. 3–19 (2018)
30. Zhang, Y., Yap, P.T., Chen, G., Lin, W., Li, W., Shen, D.: Super-resolution reconstruction of neonatal brain magnetic resonance images via residual structured sparse representation. Med. Image Anal. **55**, 76–87 (2019)
31. Zhao, L., Li, S.: Object detection algorithm based on improved YOLOv3. Electronics **9**(3), 537 (2020)

Diffusion-Based Gray-White Matter Mapping for Quantitative Tractography in Glioma Patients

Lars Smolders[1,2(✉)], Maxime Chamberland[2], Geert-Jan Rutten[1], Wouter De Baene[3], Remco van der Hofstad[2], and Luc Florack[2]

[1] Department of Neurosurgery, Elisabeth-Tweesteden Ziekenhuis, Tilburg, The Netherlands
l.smolders@etz.nl
[2] Department of Mathematics and Computer Science, Eindhoven University of Technology, Eindhoven, The Netherlands
[3] Department of Cognitive Neuropsychology, Tilburg University, Tilburg, The Netherlands

Abstract. Gliomas are malignant brain tumors whose potential to infiltrate healthy white matter (WM) necessitates a careful approach during surgical treatment. While tractography is clinically widely used in a qualitative manner to visualise WM bundles during surgery planning, quantifying the integrity of such bundles in glioma patients is a difficult challenge. Quantitative measures currently used in group-level studies are biased or heavily dependent on tractography parameters. In most populations, Anatomically Constrained Tractography (ACT) in combination with Spherical-deconvolution Informed Filtering of Tractograms (SIFT) successfully estimates WM connectivity in whole-brain tractograms. However, a critical segmentation step used in ACT typically overestimates the presence of gray matter (GM) in glioma patients due to the properties of glioma tissue on T1-weighted MRI. This leads to an incorrect derivation of the interface between GM and WM, which is used to define a seed and target region of interest (ROI) in ACT. While workarounds exist, we argue that these may bias SIFT by either under- or overestimating fiber reconstructions in and around the tumor.

In this work, we present an alternative GM-WM segmentation method based on diffusion-weighted MRI using multi-tissue Constrained Spherical Deconvolution (CSD) and derive GM-WM interfaces from these novel segmentations. We demonstrate that this method results in a more accurate GM-WM interface in and around tumors in three glioma patients, and that the resulting quantitative tractograms better identify WM disconnections induced by surgery. This method represents a step towards quantitative tractography in glioma patients, and towards improved surgical decision-making in clinical practice.

Keywords: Quantitative tractography · Glioma · Multi-tissue CSD · GM-WM interface

1 Introduction

Gliomas are malignant brain tumors that infiltrate healthy brain tissue. The first line of treatment is usually a surgical removal (resection) of as much tumor tissue as possible, in order to slow further growth and alleviate pressure on the brain. Surgeons must be careful when resecting near the boundary of the glioma, as healthy and functioning structures such as white matter (WM) bundles may be interwoven with tumor tissue. To aid in decision making, qualitative tractography in the form of bundle visualization has found widespread clinical application in surgery planning for glioma patients, as approximately knowing the extent of important fiber bundles allows surgeons to better plan paths of approach to a tumor and to avoid resecting parts of tumors that have infiltrated important bundles [18].

As an extension of this, quantitative tractography, i.e., assessing the strength of connections found by tractography, could further improve clinical decision making [20]. If one is able to quantify the strength of connections near the tumor, one could potentially make more informed decisions on whether to risk damaging these connections. For example, if a bundle that runs through the tumor is calculated to have a low connection strength, the surgeon may decide to remove this part of the tumor. In this future scenario, the oncological benefit of reducing the tumor volume outweighs the (neurological) functional impact, which is expected to be low. Despite this added utility, quantitative tractography is generally absent from clinical practice, mostly due to the lack of promising research on this subject in glioma patients. Quantitative measures that are regularly used in studies of this patient population, such as streamline counts or Fractional Anisotropy (FA), have been shown to be highly dependent on tractography parameters [6] and usually only correlate with outcomes of interest at group level [7,17]. In healthy subjects, Spherical-deconvolution Informed Filtering of Tractograms (SIFT) [12], in combination with Anatomically Constrained Tractography (ACT) [11], have been successfully used to quantify the capacity for information transfer of white matter bundles. Briefly, SIFT works by mapping a whole-brain tractogram back to voxelwise fiber densities estimated from Diffusion Weighted Imaging (DWI) data and assigning weights to each streamline based on the contribution of each streamline to the total fiber density along their paths. In order for SIFT to be correct, a whole-brain tractogram must be generated in which streamlines do not project into cortical GM, as these streamlines would be assigned higher weights than desired due to GM contributions in fiber density. ACT enforces this constraint by seeding and terminating streamlines in the GM-WM interface (GMWMI), and only allowing streamlines to pass through WM.

To find this GMWMI, the standard method supplied with ACT uses a 5-tissue segmentation based on T1-weighted MRI, where the brain is automatically segmented into WM, GM, Cerebrospinal Fluid (CSF) and deep GM (consisting of e.g. the thalamus and basal ganglia). The fifth tissue type represents pathological tissue and can be manually added. In this fifth tissue type, no assumptions regarding tractography are made, i.e., streamlines can pass through as long as the underlying diffusion data allows it. A GMWMI is then derived from the boundaries between the GM and WM segmentations. When applying this method to brains with a glioma, the problem is that glioma tissue often has GM-like intensity on T1 MRI and is consequently labeled as part of the GM by the 5-tissue segmentation. This way, spurious GMWMI is found around the boundary of the tumor, no interface is found within the tumor, and no tracking can be performed within the tumor. While a tumor segmentation can be supplied as the fifth tissue type to avoid the overestimation of GM, also no GMWMI will be found inside the tumor this way. Although many gliomas destroy GM and WM structures, it is known that especially in slowly growing Low Grade Gliomas (LGGs), functional structures may still be present within the tumor interior [8,19]. Having a poor estimate of the GMWMI within and around the glioma could result in an incorrect reconstruction of connections starting or ending in or around the tumor. Since SIFT requires an accurate whole-brain tractogram to correctly estimate the strength of each streamline, this potentially biases the connection strengths calculated by SIFT.

In this work, we addressed this issue by developing a novel procedure to derive segmentations of GM and WM, and consequently a GMWMI. This procedure is based only on DWI, using multi-tissue CSD to more accurately identify GM and WM in the tumor interior. We demonstrated that this reduced the overestimation of GMWMI around the boundary of tumors and improved the identification of GMWMI inside tumor tissue in three example glioma patients. We also demonstrated that quantitative tractography using SIFT2 [13], a more recent improvement upon the original SIFT based on the same principles, accurately identifies WM connectivity loss after surgery in one of these patients, consistent with post-operative neurological deficits experienced by this patient.

2 Methods

2.1 Patient Description

We tested our novel processing steps in three glioma patients who were treated in the Elisabeth-Tweesteden Hospital (Tilburg, The Netherlands). Two of the three patients had a Low-Grade Glioma (LGG), a slowly growing glioma that is typically less disruptive of surrounding tissue than a more aggressive High-Grade Glioma (HGG), which was found in the third patient. Representative T1 and FLAIR (Fluid-Attenuated Inversion Recovery) MRI slices are presented in Fig. 1.

- Patient 1 had two low-grade gliomas in the left frontal lobe, classified as a WHO (World Health Organization) grade II IDH-mutated (Isocitrate Dehydrogenase) astrocytoma. This patient did not suffer from post-operative neurological impairments.
- Patient 2 had a low-grade glioma in the right supplementary motor cortex, classified as a WHO grade II IDH-mutated oligodendroglioma. After surgical resection of the tumor, the patient suffered from left-sided paresis, consistent with WM damage in the tumor region.

- Patient 3 had a high-grade glioma that almost fully infiltrated the left temporal lobe, classified as WHO grade IV IDH-wild type astrocytoma. This patient did not suffer from post-operative neurological impairments.

This study was approved by the Medical Ethics Committee Brabant, The Netherlands [protocol number: NL51147.028.14]. All procedures were carried out with written informed consent of all subjects and in accordance with the principles of the Declaration of Helsinki.

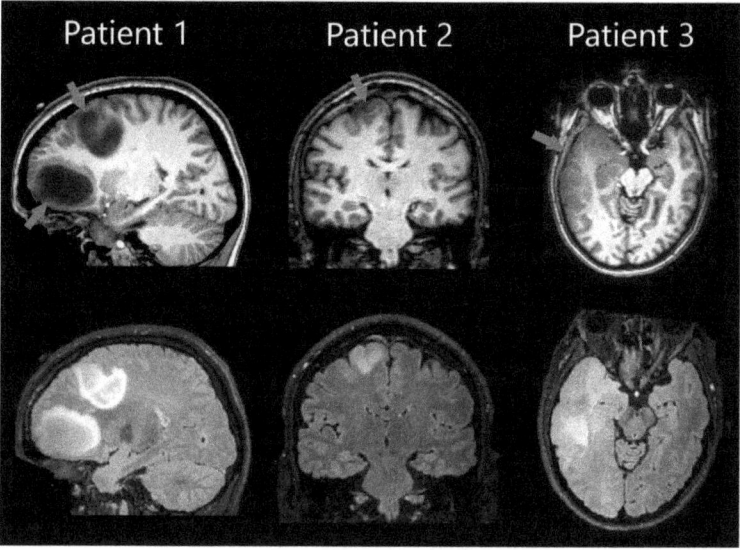

Fig. 1. Representative T1-weighted (first row) and FLAIR (second row) MRI slices of the three patients. Red arrows indicate tumor locations. Tumor-affected tissue appears as hyperintense on FLAIR imaging. (Color figure online)

2.2 Image Acquisition

Patients underwent diffusion-weighted, T1-weighted and FLAIR MRI scans three or four days before surgery and three months after surgery. MRI images were acquired using a Philips Achieva 3T MRI scanner (Philips Medical Systems, Best, The Netherlands) using a standard 32-channel radiofrequency head coil. High resolution whole-brain anatomical images were acquired using a T1-weighted sequence (TR/TE: 8.4/3.8 ms, FOV: 254 × 254 × 158 mm, flip angle: 8°, sagittal slice orientation, voxel size 1 mm isotropic). Single-shell DWI images were acquired using an echo-planar imaging (EPI) sequence (TR = 8 s; TE = 115 ms; 6 b = 0 volumes, 50 b = 1500 volumes, 2 mm isotropic voxel size), without reversed phase encoding (PE) direction.

2.3 Image Preprocessing

DWI images were denoised, corrected for distortions with top-up and eddy, and bias-corrected using MRtrix3 [15]. Since no reverse PE was acquired for the DWI in this retrospective data set, we applied non-linear registration of an inverted contrast T1 image [16] to each b0 image using ANTS [1] to most accurately register T1 images to DWI images.

In order to identify the bundles that could be damaged during surgery, we required a meaningful and accurate parcellation of the cortex. As Freesurfer usually fails processing scans of patients with large tumors [10,21], and registering an atlas from MNI (Montreal Neurological Institute) space is typically inaccurate in these cases as well, we opted to parcellate the T1 scans of patients using SLANT [4], a brain segmentation tool based on convolutional neural networks. Although this model has not been trained on brain tumor patient data, it displays a remarkable accuracy on pre-operative glioma patients in practice [9]. We observed that it does not perform well on post-operative scans, where large resection cavities may pose a problem. Therefore, we transferred the pre-operative parcellations to the post-operative DWI space by non-linear registration from pre-operative T1 to post-operative T1 and from post-operative T1 to post-operative DWI.

2.4 Novel 5-Tissue Segmentation

In order to better identify the boundary between GM and WM inside the glioma, we derived a segmentation based only on diffusion-weighted MRI using multi-tissue CSD [5]. In multi-tissue CSD, diffusion responses are divided into the three classes WM-, GM- and CSF-like, which are then used to obtain separate Fiber Orientation Distributions (FODs) for each tissue type in each voxel. The resulting FOD amplitudes can be inspected to derive the contribution of each of the three tissue types to the total diffusion signal in each voxel. Since our clinical data is acquired using only a single b-value shell, we applied Single Shell 3-Tissue CSD (SS3T-CSD) [2,3] to the DWI images, separating the FODs into contributions by WM, GM and CSF. To ensure that data were comparable between pre-operative and post-operative sessions, we averaged the response functions obtained for WM, GM and CSF between the two scanning sessions. The DWI images were segmented into WM-, GM- and CSF-dominant compartments by taking the maximum contribution voxel-wise.

We observed that parts of the prefrontal lobe WM were identified as GM-dominant by SS3T-CSD. To avoid incorrectly introducing GMWMI in this region, we sought to remove these spurious GM-like compartments. Assuming that the cortical GM is a single fully connected structure in the human brain, we replaced all GM voxels that were not directly connected to this single cortical component by WM voxels. This way, GM components that are fully enclosed by WM are automatically replaced by WM. Figure 2 illustrates the process in an example axial slice containing a spurious GM-like component. Note that this step in the process by itself is not enough to fix the issue with T1-only segmentation, as the gray-appearing glioma tissue is usually connected to the cortical GM (see Fig. 1).

Finally, the resulting WM, GM and CSF segmentations were combined with the deep-GM segmentation generated by the standard T1-based method implemented by *5ttgen* in MRtrix3 to construct a full 5-tissue segmentation image. The fifth tissue type (pathological tissue) was left empty, such that the ACT constraints were upheld throughout the entire brain. Using the 5-tissue segmentation obtained this way, we identified the boundary between the GM and WM using the standard method implemented by *5tt2gmwmi* in MRtrix3.

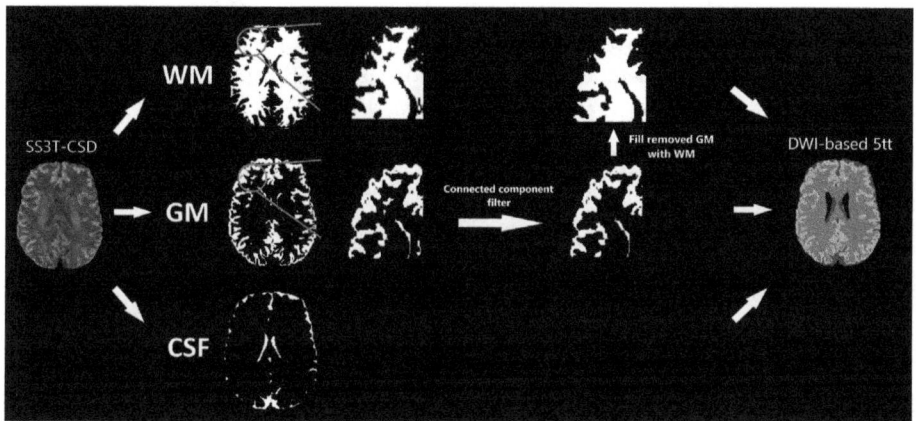

Fig. 2. Overview of the novel segmentation method. WM-, GM- and CSF-dominant voxels are identified based on multi-tissue CSD. GM-like components that are not connected to the largest GM-like component (i.e., the cortex) are replaced by WM-like voxels.

2.5 Whole-Brain Tractography

To investigate whether our novel method can correctly detect the potential WM disconnections that occurred in Patient 2 due to surgery, we applied ACT using both the standard and the alternatively generated GMWMI on both preoperative and post-operative MRI images. Tractography was performed with the iFOD2 algorithm [14] as implemented by *tckgen* in MRtrix3, using the WM FODs obtained from SS3T-CSD and supplying the custom 5-tissue segmentation image as input for ACT. Tractography parameters were: 4 million seed points seeded in the GMWMI, minimum streamline length 10 mm, maximum length 250 mm, FOD cutoff 0.06, cropping streamlines at the GMWMI, and backtracking incorrectly terminating streamlines.

Streamlines were weighted using SIFT2 [13] using the whole-brain tractograms and WM FODs as inputs. Connectivity of the ROI in Patient 2 (right supplementary motor cortex) was calculated by summing the SIFT2 weights of all streamlines between this ROI and all other SLANT parcels. Since Patient 2 suffered from left-sided paresis after surgery, we expected a large connectivity loss in this case. As controls, we calculated the same connectivity statistic for the other two patients. These two patients did not suffer any post-operative deficits, so we expected to see little connectivity loss in these cases.

3 Results

In Fig. 3 we show representative slices of the 3-tissue images as generated by SS3T-CSD, where red represents WM, green represents GM and blue represents CSF. Regions with tumor tissue display higher amounts of CSF, most likely caused by edema or demyelination effects. In Patients 2 and 3 we clearly see WM structures still present within the tumor tissue (compare with Fig. 1).

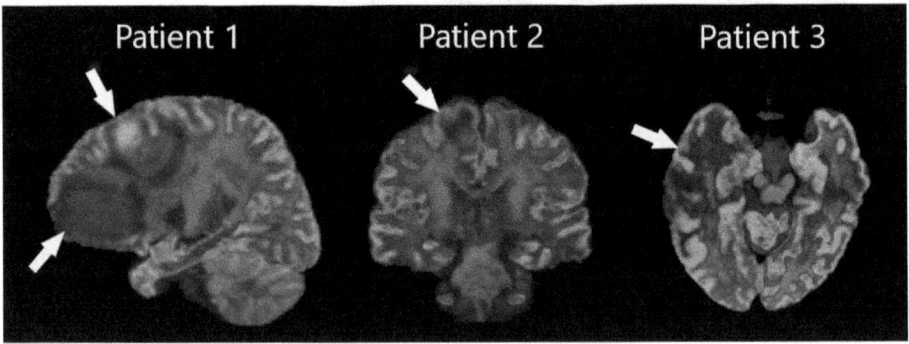

Fig. 3. 3-tissue images of the representative slices of each patient. The images are separated into white matter (red), gray matter (green) and cerebrospinal fluid (blue). White arrows indicate tumor locations. (Color figure online)

Figure 4 presents the difference between GMWMI generated from T1 MRI and GMWMI generated by our novel method in the three glioma patients. In Patient 1 we see that the T1-based GMWMI runs around the entire tumor, even though there is no GM present in these regions (see discussion). In the novel GMWMI, this boundary is not present. In other regions, the novel GMWMI mostly agrees with the T1-based GMWMI. In Patient 2 we see that the novel GMWMI penetrates further into the tumor tissue than the T1-based GMWMI, while again mostly agreeing with the T1-based GMWMI. In Patient 3, we again see more GMWMI inside the tumor.

Figure 5 shows the difference in connectivity (i.e., summed SIFT2 weights) calculated by the two methods on Patient 2. Considering the right supplementary motor cortex, we see that the novel method calculates a connectivity loss of −63, going from 101 to 38, while the standard method calculates only −20, going from 69 to 49.

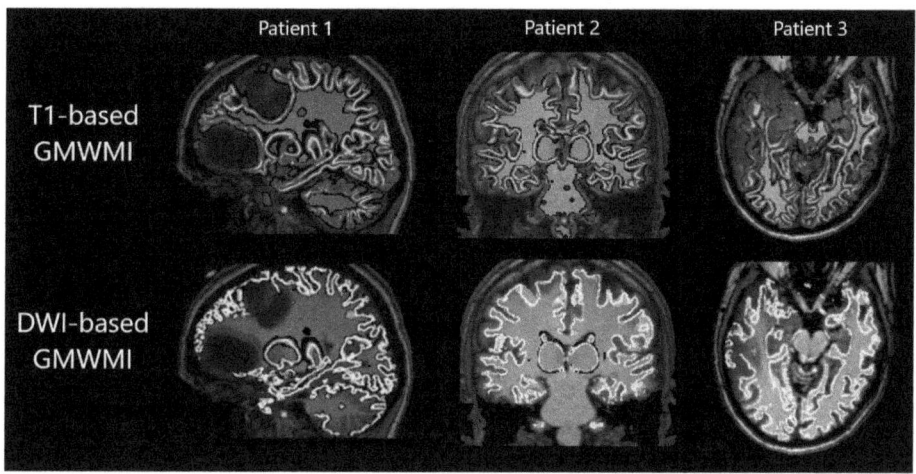

Fig. 4. Comparison of GMWMI as generated from T1-based *5ttgen* and generated from multi-tissue CSD on each of the three patients.

In Fig. 6 the same comparison is shown for Patient 1 and Patient 3 respectively. In Patient 1, the novel method calculates a connectivity loss of −4, going from 85 to 81, while the standard method calculates a connectivity gain of +10, going from 73 to 83. In Patient 3, the novel method calculates a connectivity loss of −5, going from 78 to 73, while the standard method calculates a connectivity loss of −3, going from 71 to 68.

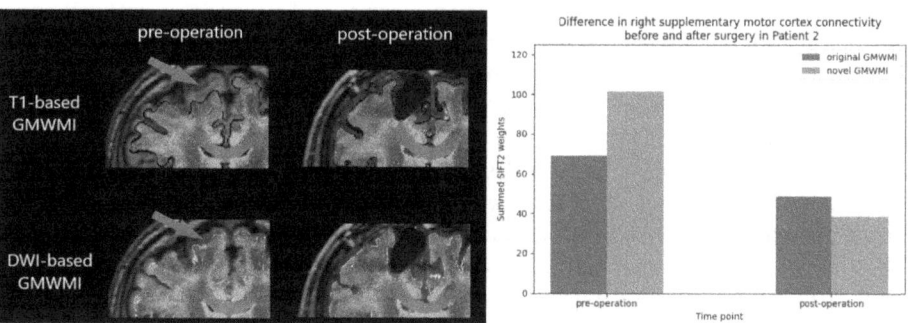

Fig. 5. Difference in connectivity calculation between the standard method and our novel method on Patient 2. The novel method finds stronger connections to the right supplementary motor cortex before surgery than the standard method, and consequently measures a larger loss of connectivity after surgery.

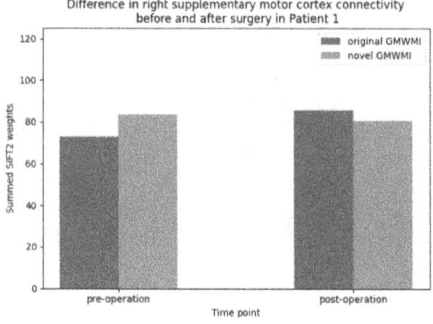 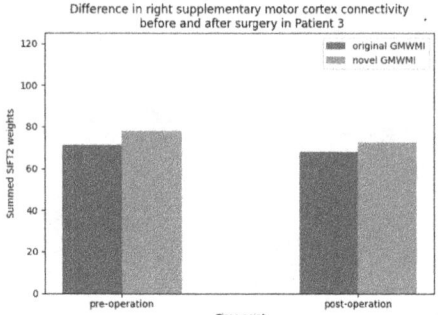

Fig. 6. Difference in connectivity calculation between the standard method and our novel method on Patients 1 and 3. Both methods find minimal changes in connectivity before and after surgery in the right supplementary motor cortex.

4 Discussion

Our results indicate that our novel DWI-based method of segmenting MRI into GM and WM is better able to identify GMWMI in and around gliomas. This improvement allowed us to more accurately construct a whole-brain tractogram, and thus better quantify the integrity of important bundles in and around gliomas, than the standard T1-based GMWMI. In Patient 2, we see in Fig. 3 that there is some WM structure present inside the tumor. The new method uses this information to fix the underestimation of GMWMI inside tumor tissue (Fig. 4), and consequently finds a clear disconnection of the WM bundles passing through the tumor (Fig. 5) as a result of surgery. This is consistent with the left-sided paresis suffered by this patient after surgery and suggests an improvement over the standard method, which detects only a slight decrease in connectivity. This large disconnection is not found in the other two patients by our novel method (Fig. 6), which is consistent with the absence of post-operative deficits in these patients.

Furthermore, the new method fixes the overestimation of GMWMI around the tumors in Patient 1. In Fig. 3 we see that most of the volume in the two tumors is identified as CSF-dominant (blue). While on the FLAIR image in Fig. 1 we can clearly see that the tumor volumes do not consist of only free water (as CSF or excessive edema would be fully black on FLAIR), we can infer from the combination of T1, FLAIR and SS3T images that the tumor should not have internal WM or GM structures. Furthermore, we do not expect any cortical GM to be present in the deep WM areas, relatively far away from the cortex. From this we infer that there should be no GM on the entire boundary of these tumors in which streamlines should start or terminate. This is correctly identified by the new method, but not by the standard method, which incorrectly infers from the lower T1 intensity observable in Fig. 1 that there should be GM.

In conclusion, our new method provides an alternative approach to constructing GMWMI in glioma patients, which can be used to better quantitatively measure WM disconnection due to surgery. If expanded upon in future work, our methods provide a step-stone for additional research in quantitative tractography for glioma patients.

4.1 Limitations

The novel GMWMI construction method depends on the multi-tissue CSD algorithm used and the quality of the acquired DWI data. As is clear from the overestimated GM in prefrontal regions observed in Patient 3, clinical single-shell DWI of limited quality poses some issues for SS3T-CSD, potentially due to noise in the DWI acquisitions. We have tried to alleviate this issue by removing GM that is not connected to the largest connected GM component. However, if there is too much overestimated GM and this overestimated GM is connected to the main cortical component, this does not work sufficiently well, as shown in Fig. 7.

Considering the disconnection results presented in Fig. 5, it is possible that inaccurate registration between the T1-weighted image and the diffusion-weighted images slightly misplaced the T1-based GMWMI and thus caused mistakes in seeding and terminating streamlines. However, we visually did not observe any major registration errors between the two modalities and expect the effects of such errors to be small. Furthermore, the tumors of Patients 1 and 3 posed larger problems for the registration algorithm than the tumor of Patient 2, but we only found minor connectivity differences between the two methods in these cases (Fig. 6).

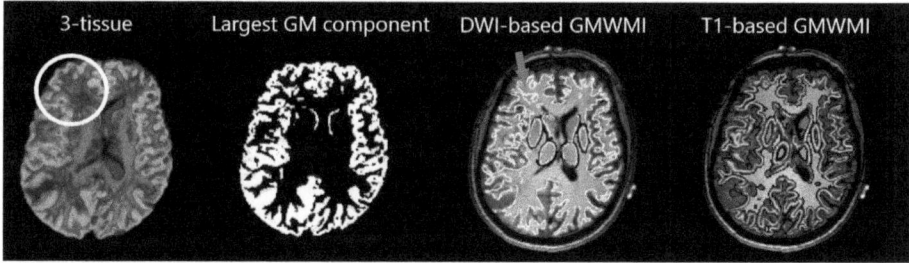

Fig. 7. Demonstration of overestimation of GM in prefrontal regions. The spurious GM is connected to the largest component and is thus not removed by our method, consequently generating spurious GMWMI in the area. The T1-based GMWMI does not have this issue.

4.2 Future Work

Our method could best be further improved by removing the overestimated GM in the prefrontal regions. Improvements to multi-tissue CSD (or other methods of segmenting GM and WM based on DWI) could automatically solve this issue. Alternatively, removing these spurious GM-like components using methods based on anatomical priors or geometric information could alleviate this problem.

More extensive quantitative studies should be performed to investigate the descriptive and predictive potential of our approach. For example, relating the post-operative deficits of patients to disconnections detected by this method, as we have done for a single patient in this work, should be extended to a larger group of patients and statistically compared to existing methods. Specifically, the prevalence of spurious GM compartments connected to the main cortical GM, as described in Sect. 4.1, should be quantitatively investigated. If successful, such a study could reveal whether quantitative tractography incorporating our method is capable of detecting disconnections due to surgery, which in the future could assist in neurosurgical decision-making.

Acknowledgments. This publication is part of the project "A personalized care path for brain tumor patients", (partly) funded by The Netherlands Organisation for Health Research and Development (ZonMw), project number 10070012010006, and of the project "Bringing Tractography into Daily Neurosurgical Practice", (partly) financed by the Dutch Research Council (NWO), project number KICH1.ST03.21.004. The work of Remco van der Hofstad is supported in part by the Netherlands Organisation for Scientific Research (NWO) through the Gravitation NETWORKS grant no. 024.002.003.

Disclosure of Interests. The authors have no competing interests to declare that are relevant to the content of this article.

References

1. Avants, B.B., Tustison, N.J., Song, G., Cook, P.A., Klein, A., Gee, J.C.: A reproducible evaluation of ANTs similarity metric performance in brain image registration. Neuroimage **54**, 20332044 (2011)
2. Dhollander, T., Connelly, A.: A novel iterative approach to reap the benefits of multi-tissue CSD from just single-shell (b=0) diffusion MRI data. In: 24th International Society of Magnetic Resonance in Medicine, vol. 24, p. 3010 (2016)
3. Dhollander, T., Mito, R., Raffelt, D., Connelly, A.: Improved white matter response function estimation for 3-tissue constrained spherical deconvolution. In: 27th International Society of Magnetic Resonance in Medicine, vol. 27, p. 555 (2019)
4. Huo, Y., et al.: 3D whole brain segmentation using spatially localized atlas network tiles. Neuroimage **194**, 105–119 (2019)
5. Jeurissen, B., Tournier, J.D., Dhollander, T., Connelly, A., Sijbers, J.: Multi-tissue constrained spherical deconvolution for improved analysis of multi-shell diffusion MRI data. Neuroimage **103**, 411426 (2014)
6. Jones, D., Knösche, T., Turner, R.: White matter integrity, fiber count, and other fallacies: the do's and don'ts of diffusion MRI. Neuroimage **73**, 239–254 (2013)
7. Jütten, K., et al.: Dissociation of structural and functional connectomic coherence in glioma patients. Sci. Rep. **11** (2021)
8. Mato, D., Velasquez, C., Gómez, E., Marco de Lucas, E., Martino, J.: Predicting the extent of resection in low-grade glioma by using intratumoral tractography to detect eloquent fascicles within the tumor. Neurosurgery **88(2)**, 190202 (2021)

9. Meesters, S., Landers, M., Rutten, G., Florack, L.: Subject-specific automatic reconstruction of white matter tracts. J. Digit. Imaging **36**, 26482661 (2023)
10. Radwan, A.M., et al.: Virtual brain grafting: enabling whole brain parcellation in the presence of large lesions. Neuroimage **229**, 117731 (2021)
11. Smith, R.E., Tournier, J.D., Calamante, F., Connelly, A.: Anatomically-constrained tractography: improved diffusion MRI streamlines tractography through effective use of anatomical information. Neuroimage **62**(3), 1924–1938 (2012)
12. Smith, R.E., Tournier, J.D., Calamante, F., Connelly, A.: SIFT: spherical-deconvolution informed filtering of tractograms. Neuroimage **67**, 298–312 (2013)
13. Smith, R.E., Tournier, J.D., Calamante, F., Connelly, A.: SIFT2: enabling dense quantitative assessment of brain white matter connectivity using streamlines tractography. Neuroimage **119**, 338–351 (2015)
14. Tournier, J.D., Calamante, F., Connelly, A.: Improved probabilistic streamlines tractography by 2nd order integration over fibre orientation distributions. In: Proceedings of the International Society for Magnetic Resonance in Medicine, p. 1670 (2010)
15. Tournier, J.D., et al.: MRtrix3: a fast, flexible and open software framework for medical image processing and visualisation. Neuroimage **202**, 116137 (2019)
16. Wang, S., Peterson, D.J., Gatenby, J.C., Li, W., Grabowski, T.J., Madhyastha, T.M.: Evaluation of field map and nonlinear registration methods for correction of susceptibility artifacts in diffusion MRI. Front. Neuroinform. **11** (2017)
17. Wei, Y., et al.: Structural connectome quantifies Tumour invasion and predicts survival in glioblastoma patients. Brain **146**(4), 1714–1727 (2022)
18. Yang, J.Y.M., Yeh, C.H., Poupon, C., Calamante, F.: Diffusion MRI tractography for neurosurgery: the basics, current state, technical reliability and challenges. Phys. Med. Biol. **66** (2021)
19. Yeh, F., Irimia, A., Bastos, D., Golby, A.: Tractography methods and findings in brain tumors and traumatic brain injury. Neuroimage **15** (2021)
20. Zhang, F., et al.: Quantitative mapping of the brains structural connectivity using diffusion MRI tractography: a review. NeuroImage **249**, 118870 (2022)
21. Zhang, F., et al.: Automated connectivity-based groupwise cortical atlas generation: application to data of neurosurgical patients with brain tumors for cortical parcellation prediction. In: 2017 IEEE 14th International Symposium on Biomedical Imaging (ISBI 2017), pp. 774–777 (2017)

Ground-Truth Effects in Learning-Based Fiber Orientation Distribution Estimation in Neonatal Brains

Rizhong Lin[1,2,3], Hamza Kebiri[2,4(✉)], Ali Gholipour[5,6], Yufei Chen[3], Jean-Philippe Thiran[1,2,4], Davood Karimi[5], and Meritxell Bach Cuadra[2,4]

[1] Signal Processing Laboratory (LTS5), École Polytechnique Fédérale de Lausanne (EPFL), Lausanne, Switzerland
[2] Department of Radiology, Lausanne University Hospital (CHUV) and University of Lausanne (UNIL), Lausanne, Switzerland
hamza.kebiri@unil.ch
[3] College of Electronic and Information Engineering, Tongji University, Shanghai, China
[4] CIBM Center for Biomedical Imaging, Lausanne, Switzerland
[5] Computational Radiology Laboratory, Department of Radiology, Boston Children's Hospital and Harvard Medical School, Boston, MA, USA
[6] Department of Radiological Sciences, University of California Irvine, Irvine, CA, USA

Abstract. Diffusion Magnetic Resonance Imaging (dMRI) is a non-invasive method for depicting brain microstructure *in vivo*. Fiber orientation distributions (FODs) are mathematical representations extensively used to map white matter fiber configurations. Recently, FOD estimation with deep neural networks has seen growing success, in particular, those of neonates estimated with fewer diffusion measurements. These methods are mostly trained on target FODs reconstructed with *multi-shell multi-tissue constrained spherical deconvolution* (MSMT-CSD), which might not be the ideal ground truth for developing brains. Here, we investigate this hypothesis by training a state-of-the-art model based on the U-Net architecture on both MSMT-CSD and *single-shell three-tissue constrained spherical deconvolution* (SS3T-CSD). Our results suggest that SS3T-CSD might be more suited for neonatal brains, given that the ratio between single and multiple fiber-estimated voxels with SS3T-CSD is more realistic compared to MSMT-CSD. Additionally, increasing the number of input gradient directions significantly improves performance with SS3T-CSD over MSMT-CSD. Finally, in an age domain-shift setting, SS3T-CSD maintains robust performance across age groups, indicating its potential for more accurate neonatal brain imaging.

Keywords: FOD estimation · Neonatal brain · Deep learning · SS3T-CSD · MSMT-CSD · Age domain shift

R. Lin and H. Kebiri—Equal contribution.

1 Introduction

Diffusion Magnetic Resonance Imaging (dMRI) is a crucial tool for analyzing *in vivo* the microstructural organization of white matter fibers in the brain. This technique measures the random motion of water molecules, providing unique insights into brain connectivity and revealing abnormalities undetectable by other modalities both *in vivo* and non-invasively. The accurate depiction of white matter (WM) fibers using dMRI is particularly important in early brain development stages, where WM is still in maturation. Advanced methods, such as constrained spherical deconvolution (CSD) [30] and multi-shell multi-tissue CSD (MSMT-CSD) [12], are used to reconstruct orientation distribution functions (FODs), enabling both local and global quantitative analyses [4].

Supervised deep learning on high-quality datasets have enabled a growing number of methods [11,14–16,20,21,26,28] to predict FODs from the raw diffusion signal [11,15,28] or its spherical harmonics (SH) representation [14,16,20,21], or from a spherical deconvolution model [26]. Some of these methods [15,17] have been successfully applied to developing brains for which acquisition times are more constrained, and hence a lower number of measurements is available. In particular, [17] employed a U-Net architecture to predict large patches of FODs from the newborns of the Developing Human Connectome Project (dHCP), using as few as six diffusion measurements and one b0 as input, making it suitable for clinical acquisitions; [19] investigated age- and site-related domain shifts for fiber estimation within rapidly developing populations. The vast majority of these methods either focus on innovating at the architecture level, the learning choices and parameters, or the input format to the network.

The target ground truth (GT) used to be learned from is not often discussed. The majority of the works [11,14–16,20,26,28] use MSMT-CSD. However, its use for developing brains has been questioned [23] given the similarity of the response function profiles for each b-value and each tissue at this maturation stage. In fact, the mean signal across b-values of the gray matter (GM) lies within the WM intensities, as opposed to adult data where it lies below, allowing a clear distinction between tissues [23]. This distinction between tissues is also age-dependent, making the problem of estimating accurate FODs even more challenging for neonates. Recently, [7] has shown that employing the b1000 shell in the case of newborn (dHCP) data with a method relying on a single non-zero b-value measurement and a b0[1], named single-shell three-tissue CSD (SS3T-CSD) [5,6], offers superior detection of crossing fibers compared to MSMT-CSD using all the available shells (b400, b1000 and b2600).

In this work, we explored the suitability of SS3T-CSD for deep learning-based FOD estimation in early development stages. We began by evaluating the consistency of GTs generated by MSMT-CSD, commonly used in this context, and SS3T-CSD, which we investigated here. Next, we extensively tested the state-of-the-art deep learning FOD estimation method on 207 newborn subjects, assessing the impact of training with SS3T-CSD compared to MSMT-CSD GT

---

[1] bx denotes diffusion weighting, where x is the b-value represented in s/mm^2.

for the first time. Finally, we examined potential domain shift effects due to age by comparing the performance of models trained and tested on cohorts of similar and different ages using both SS3T-CSD and MSMT-CSD GTs.

2 Methodology

2.1 Datasets and Processing

We used neonatal brain MRI data from the 3rd release of the Developing Human Connectome Project (dHCP) [8], which consists of newborn subjects scanned at the post-menstrual age (PMA) of 26 to 45 weeks using a 3T Philips Achieva scanner. The dMRI data release, having a multi-shell sequence with b-values $0, 400, 1000$ and 2600 s/mm^2 and 20, 64, 88 and 128 measurements respectively, had been preprocessed using SHARD [3]. The resulting resolution is $1.5 \times 1.5 \times 1.5$ mm^3, covering a field of view of $100 \times 100 \times 64$ voxels.

Following the approach in [16], we generated a WM mask by integrating the *White Matter* and *Brainstem* labels from the provided parcellation of the T2-weighted image of each subject, aligning it to the space of the respective dMRI image using ITK-SNAP [31]. Voxels with Fractional Anisotropy (FA) values (computed using MRtrix [29]) greater than 0.25 were also included.

For our experiments (detailed in Sect. 2.3), we selected two subsets, denoted as S_1 and S_2, totaling 207 unique subjects:

- **Subset S_1**: 100 subjects (PMA at scan: [35.57, 44.29], median: 40.86, mean: 40.11, SD: 2.38).
- **Subset S_2** divided into two age groups:
  - *Early-stage group* (S_{2e}): 65 subjects (PMA at scan: [33.29, 37.86], median: 35.57, mean: 35.69, SD: 1.41).
  - *Late-stage group* (S_{2l}): 65 subjects (PMA at scan: [41.0, 45.14], median: 42.29, mean: 42.40, SD: 1.07).

Each subject's data preparation for the deep learning pipeline (described in Sect. 2.2) involves sampling single-shell b1000 dMRI images from the full sequence as proposed in [27]. These images are normalized by a single b0 image to reduce b-value dependency and then projected onto the corresponding SH spaces to enhance acquisition independence.

2.2 FOD Estimation Learning-Based Model

The method [17] is based on learning a mapping between the SH representation of the raw diffusion signal and the FOD (Fig. 1, top).

Ground-Truth Models. MSMT-CSD extends (single-shell single-tissue) CSD [30] to accommodate multiple shells and multiple tissues; it aims to solve a constrained linear least squares problem using convex quadratic programming [12], in order to find the optimal FOD coefficients for each tissue, requiring acquisitions with at least 3 b-values. On the other hand, SS3T-CSD addresses the optimization problem of finding the optimal FOD coefficients differently: initially, it fixes prior WM coefficients and estimates GM and cerebrospinal fluid (CSF) coefficients; then, it fixes the CSF coefficients and estimates WM and GM coefficients—this two-step iteration is repeated until convergence [5], allowing the method to rely on only two b-value samples (non-b0 and b0).

Network Specifications. The network receives a $16 \times 16 \times 16$ patch of SH representations from n_{sig} single-shell diffusion measurements (b1000) as input and outputs the corresponding patch of FODs represented in the same SH basis (SH-L_{max} order 8). We train two models: one targeting FODs estimated from all 300 multi-shell measurements using MSMT-CSD [12], and the other from 88 $b1000$ and 20 $b0$ measurements using SS3T-CSD [5], as depicted upper in Fig. 1.

Training. For each experimental setting, training was conducted using the Adam optimizer [18] to minimize the ℓ_2 norm loss of the predicted 45 SH coefficients against the GT FOD SH coefficients, which were generated with MRtrix [29] or one of its forks, MRtrix3Tissue (https://3Tissue.github.io).

During each epoch, 128 patches were randomly extracted from non-empty volumes of each training subject, with each batch containing all 128 patches at a batch size of 1. The training regimen featured an initial learning rate of 1×10^{-4}, a dropout rate of 0.1, and used an early stopping strategy. A sliding window technique was employed during inference to process all patches consecutively.

Implementation and Code Availability. Training was performed on an NVIDIA RTX 2080 Ti GPU, each session lasting approximately 24 h, using PyTorch [22], Lightning [9], and MONAI [2]. The code is available at https:// github.com/Medical-Image-Analysis-Laboratory/dl_fiber_domain_shift.git.

2.3 Experiments

Experiments testing the efficacy of deep learning models in predicting FODs from MSMT- and SS3T-CSD using dMRI data are outlined in Fig. 1.

(i) Ground-Truth Consistency. Following the methodology described in [17], we assess GT consistency using MSMT-CSD and SS3T-CSD algorithms on two distinct half subsets of dMRI measurements: 150 measurements each for MSMT-CSD (10 b0's, 32 b400's, 44 b1000's, and 64 b2600's) and 54 measurements each for SS3T-CSD (10 b0's and 44 b1000's). This method evaluates the reliability of the GT algorithms by comparing FODs generated from equivalent but separate subsets, using 80 subjects randomly sampled from data subset S_1.

(ii) Model Assessment and Ablation Study on the Number of Input Directions. Performance between models trained on SS3T-CSD and those trained on MSMT-CSD was compared. This comparison was also performed across a different number of input directions to the network. The network input

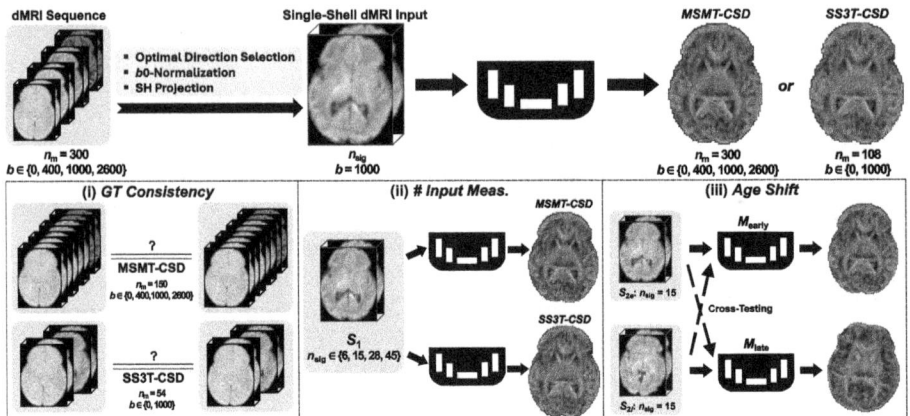

Fig. 1. Workflow overview. Single-shell b1000 dMRI images are sampled from the full series, normalized by b0 and projected into SH space. These images are inputs to a deep learning model predicting GT FODs, using either MSMT-CSD (300 meas.) or SS3T-CSD (108 meas.). The experiments assess: (i) consistency of GT algorithms, (ii) impact of input quantity on model performance, and (iii) model effectiveness across different neonatal developmental stages. Illustrative brain images are from dHCP.

has n_{sig} measurements ($n_{\text{sig}} \in \{6, 15, 28, 45\}$) from the b1000 shell, each projected onto an SH basis with $L_{\max} \in \{2, 4, 6, 8\}$. In this phase, the data subset S_1 was divided into groups of 70 for training, 15 for validation, and 15 for testing.

(iii) Age-Related Domain Shifts. To explore age-related variations within each dataset, we conducted age-specific training and testing within and across these ages [19]. As mentioned in Sect. 2.1, in the dHCP dataset, subjects in S2 were selected and divided into two age groups, denoted as *early* and *late*, respectively. Each age group consists of 65 subjects, further partitioned into splits of 40 for training, 10 for validation, and 15 for testing. We fixed the number of input measurements to 15 (corresponding to SH order 4) and trained separate models on the *early* and *late* age groups. Each model was then tested on its own age group (self-testing) as well as on the other age group (cross-age testing).

2.4 Evaluation Metrics

Quantitative validation was performed based on the agreement rate (AR) [17] between the number of fibers estimated by the number of peaks extracted using Dipy [10] (with a mean separation angle of 45°, a maximum number of 3 peaks and relative peak threshold of 0.5), the angular error (AE) between those peaks [17], and the apparent fiber density (AFD) [24] error. For the AFD error, we report the mean absolute percentage error (MAPE) as the magnitude of the SS3T-CSD-derived FODs is higher than that of MSMT-CSD.

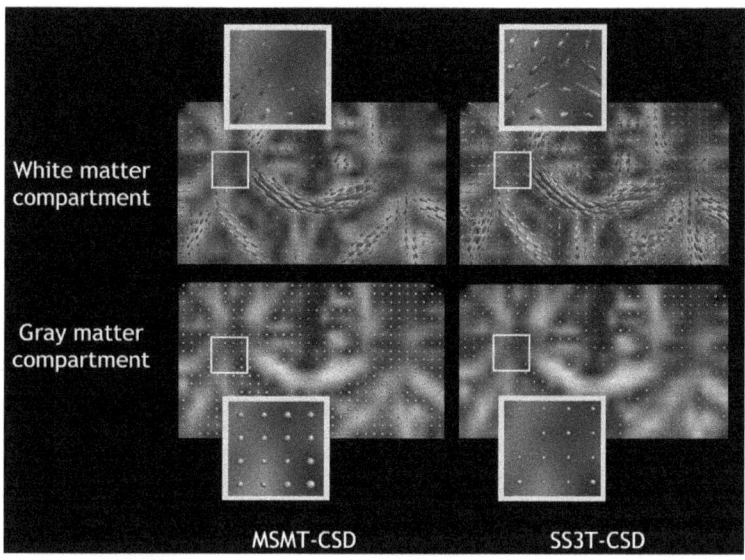

Fig. 2. Qualitative comparison of coronal slices of the FODs used as GT for the model, reconstructed with MSMT-CSD and SS3T-CSD, respectively, from the dMRI scan of a subject at 40 weeks PMA. Both gray matter and white matter compartments are displayed on the fractional anisotropy (FA) image computed from dMRI data. FOD estimation and visualization were performed with MRtrix [29].

3 Results

Firstly, we show a qualitative comparison between the FOD reconstructed using the two GT methods for a neonate with 40 weeks PMA (Fig. 2), where we clearly see missing crossings in MSMT-CSD, likely due to the overestimation of the gray matter compartment, compared to SS3T-CSD.

3.1 Ground-Truth Consistency

Table 1 provides a comparison between the two GT methods. MSMT-CSD shows a higher agreement rate for single fibers but similar angular errors when compared to SS3T-CSD. In contrast, in the 2-fiber configuration, SS3T-CSD outperforms MSMT-CSD in agreement rate and exhibits slightly better angular errors. It is fair to mention that SS3T-CSD employs the minimum number of directions given the SH-order of the FOD (44 directions for 45 coefficients), compared to 140 directions by MSMT-CSD. Hence, the SS3T-CSD consistency errors reported here represent the upper bounds of expected errors. This is particularly relevant for 3-fiber configurations, which are more susceptible to noise and thus less reliable. Importantly, closer examination of the confusion matrices (Table 2) reveals that the proportion of multiple fibers estimated by MSMT-CSD is around 23%, significantly lower than the literature-reported values of over 60% [13,25], whereas SS3T-CSD estimates this proportion to be 61%.

Table 1. Quantitative comparison of the consistency of GT FODs generated with MSMT-CSD and SS3T-CSD. Agreement Rate (AR, %) and Angular Error (AE, °) for each fiber number configuration, and AFD MAPE (ΔAFD, %) are listed, together with the number of measurements (n_m) used and the corresponding b-values. AE is computed among fibers with agreed peak predictions only.

Method	b-values	n_m	1-Fiber		2-Fiber		3-Fiber		ΔAFD
			AR	AE	AR	AE	AR	AE	
MSMT-CSD	$\{0, 400, 1000, 2600\}$	150	88.8	6.92	45.6	14.91	52.8	25.96	2.87
SS3T-CSD	$\{0, 1000\}$	54	63.9	6.76	57.7	12.32	42.1	23.38	0.43

Table 2. Confusion matrices in the number of peaks agreement, normalized over all populations (in %) for MSMT-CSD and SS3T-CSD ground truth consistency.

MSMT-CSD

# Fibers	1	2	3
1	0.715	0.0446	0.0052
2	0.0362	0.1013	0.021
3	0.0035	0.0188	0.0544

SS3T-CSD

# Fibers	1	2	3
1	0.2955	0.0775	0.0062
2	0.0774	0.3517	0.0494
3	0.0056	0.0532	0.0835

3.2 Model Assessment and Number of Input Directions Effect

Figure 3 illustrates the comparative results in terms of agreement rate, angular error, and AFD error across varying numbers of input gradient directions, using MSMT-CSD and SS3T-CSD GTs, respectively.

We observe that SS3T-CSD consistently shows higher agreement rates and lower angular errors for multiple fibers compared to MSMT-CSD across all input directions, except for 6 directions where they are comparable. Increasing the number of input directions improves the network's accuracy in predicting multiple fibers more significantly with SS3T-CSD than MSMT-CSD. Specifically, SS3T-CSD achieves an agreement rate of around 70%–76% for single and 2-fiber populations with 28–45 input directions, while MSMT-CSD's 2-fiber population agreement rate remains around 30% compared to 88% for single fibers, even with 45 input directions. However, for single fiber populations, MSMT-CSD produces more accurate predictions. For AFD, SS3T-CSD shows lower errors across all input levels compared to MSMT-CSD, with the former exhibiting lower error variance. This can be attributed to the fact that AFD is computed using the same b-value as the input b-value measurement to the network for SS3T-CSD, whereas it is computed with all three shells in MSMT-CSD.

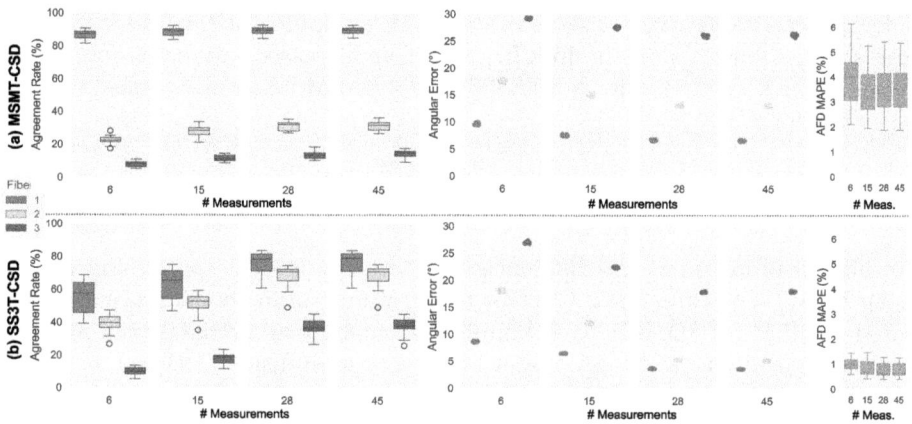

Fig. 3. Comparison of performance of the models based on incrementing numbers of input directions, using GT: (a) MSMT-CSD; (b) SS3T-CSD. AR and AE under different fiber number configurations and AFD Error are depicted.

3.3 Age Domain Shift

Figure 4 presents the effects of age-specific training and testing on model performance using MSMT- and SS3T-CSD GTs. We first notice that when testing in the *early*-stage cohort, SS3T-CSD cross-age training is more robust to domain shift across all metrics, including single-fiber populations AR and AE, compared to MSMT-CSD. In fact, the drop for instance in 2-fibers agreement rate goes from around 28% to 18% for the latter, but only from 55% to 50% for the former. However, for the configuration when tested in the *late*-stage cohort, no significant

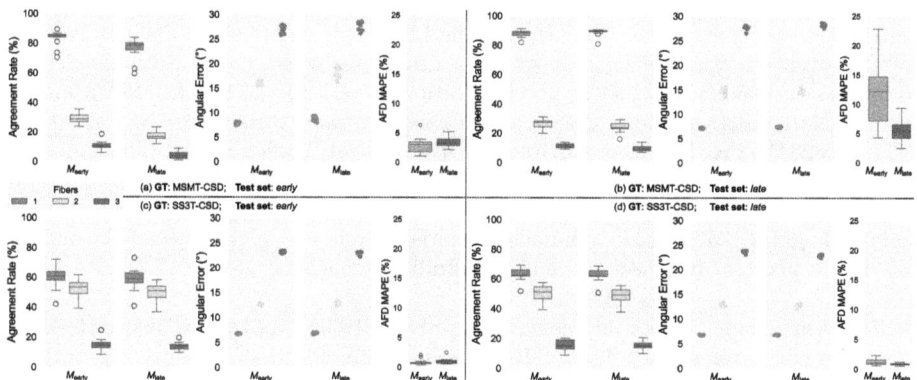

Fig. 4. Comparison of performance metrics for FOD estimation models trained and tested on *early* and *late* developmental stages using MSMT-CSD and SS3T-CSD ground truths. The agreement rate, angular error, and AFD error are depicted across models trained on *early* (M_{early}) and *late* (M_{late}) age groups.

difference can be observed between the two GTs, except for AFD where SS3T-CSD seems more robust to domain shift. In general, an increased variance is observed in cross-testing scenarios in for both SS3T-CSD- and MSMT-CSD-trained models, especially for AFD.

4 Conclusion

We have demonstrated the differences in the training of deep learning models using MSMT-CSD and SS3T-CSD for FOD estimation in neonatal brains. Compared to MSMT-CSD, SS3T-CSD improved accuracy and reliability in multiple fiber configurations, which is a major bottleneck in diffusion MRI and, in particular, for rapidly developing brains [1]. However, there is a drop in performance in single-fiber populations compared to MSMT-CSD, potentially due to higher values in gray matter compartments in GT MSMT-CSD that smoothens the overall estimated FODs, resulting in less predicted multi-fibers voxels, and vice-versa for SS3T-CSD and multiple fibers predictions. More b1000 measurements for SS3T-CSD can potentially lower this effect. Furthermore, SS3T-CSD showed robust performance across different age groups, reducing the performance drop caused by age domain shifts.

Future work will focus on evaluating SS3T-CSD in other pediatric populations, such as fetuses and babies, and exploring downstream performance in tractography. We will also investigate a variety of models beyond U-Net and different strategies to address domain shift challenges through data harmonization and domain adaptation techniques.

Acknowledgments. We acknowledge access to the facilities and expertise of the CIBM Center for Biomedical Imaging, a Swiss research center of excellence founded and supported by Lausanne University Hospital (CHUV), University of Lausanne (UNIL), École Polytechnique Fédérale de Lausanne (EPFL), University of Geneva (UNIGE), Geneva University Hospitals (HUG), and the Leenaards and Jeantet Foundations.

This research was partly supported by grants from the Swiss National Science Foundation (182602 and 215641); the US National Institutes of Health (NIH), including: the National Institute of Neurological Disorders and Stroke (R01NS106030 and R01NS128281), the National Institute of Biomedical Imaging and Bioengineering (R01EB032366), and the Eunice Kennedy Shriver National Institute of Child Health and Human Development (R01HD110772); and the National Natural Science Foundation of China (62173252). The opinions expressed herein are solely those of the authors and do not necessarily reflect those of the funding agencies.

Ethics Statement. This study uses pre-approved dHCP data by UK Health Research Authority (REC reference: 14/LO/1169), requiring no additional ethical approval.

Disclosure of Interests. The authors declare no relevant competing interests regarding the content of this article.

References

1. Calixto, C., et al.: White matter tract crossing and bottleneck regions in the fetal brain. bioRxiv 2024.07.20.603804 (2024)
2. Cardoso, M.J., et al.: MONAI: an open-source framework for deep learning in healthcare. arXiv preprint arXiv:2211.02701 (2022)
3. Christiaens, D., et al.: Scattered slice SHARD reconstruction for motion correction in multi-shell diffusion MRI. Neuroimage **225**, 117437 (2021)
4. Descoteaux, M.: High angular resolution diffusion MRI: from local estimation to segmentation and tractography. PhD thesis, Université Nice Sophia Antipolis (2008)
5. Dhollander, T., Connelly, A.: A novel iterative approach to reap the benefits of multi-tissue CSD from just single-shell (+b=0) diffusion MRI data. In: ISMRM, p. 3010 (2016)
6. Dhollander, T., Raffelt, D., Connelly, A.: Unsupervised 3-tissue response function estimation from single-shell or multi-shell diffusion MR data without a co-registered T1 image. In: ISMRM Workshop on Breaking the Barriers of Diffusion MRI (2016)
7. Dhollander, T., et al.: Improved white matter response function estimation for 3-tissue constrained spherical deconvolution. In: ISMRM, p. 0555 (2019)
8. Edwards, A.D., et al.: The Developing human connectome project neonatal data release. Front. Neurosci. **16** (2022)
9. Falcon, W., The PyTorch Lightning team, PyTorch Lightning (2024). https://doi.org/10.5281/zenodo.11644096
10. Garyfallidis, E., et al.: Dipy, a library for the analysis of diffusion MRI data. Front. Neuroinform. **8** (2014)
11. Hosseini, S., et al.: CTtrack: a CNN+Transformer-based framework for fiber orientation estimation & tractography. Neurosci. Inform. **2**(4), 100099 (2022)
12. Jeurissen, B., et al.: Multi-tissue constrained spherical deconvolution for improved analysis of multi-shell diffusion MRI data. Neuroimage **103**, 411–426 (2014)
13. Jeurissen, B., et al.: Investigating the prevalence of complex fiber configurations in white matter tissue with diffusion magnetic resonance imaging. Hum. Brain Mapp. **34**(11), 2747–2766 (2013)
14. Jha, R.R., et al.: Undersampled single-shell to MSMT fODF reconstruction using CNN-based ODE solver. Comput. Methods Programs Biomed. **230**, 107339 (2023)
15. Karimi, D., et al.: Learning to estimate the fiber orientation distribution function from diffusion-weighted MRI. Neuroimage **239**, 118316 (2021)
16. Kebiri, H., et al.: Deep learning microstructure estimation of developing brains from diffusion MRI: a newborn and fetal study. Med. Image Anal. **95**, 103186 (2024)
17. Kebiri, H., et al.: Robust estimation of the microstructure of the early developing brain using deep learning. In: MICCAI, pp. 293–303 (2023)
18. Kingma, D.P., Ba, J.: Adam: a method for stochastic optimization. In: ICLR (2015)
19. Lin, R., et al.: Cross-age and cross-site domain shift impacts on deep learning-based white matter fiber estimation in newborn and baby brains. In: IEEE ISBI (2024)
20. Lin, Z., et al.: Fast learning of fiber orientation distribution function for MR tractography using convolutional neural network. Med. Phys. **46**(7), 3101–3116 (2019)
21. Nath, V., et al.: Deep learning reveals untapped information for local white-matter fiber reconstruction in diffusion-weighted MRI. Magn. Reson. Imaging **62**, 220–227 (2019)

22. Paszke, A., et al.: PyTorch: an imperative style, high-performance deep learning library. In: Advances in Neural Information Processing Systems. Curran Associates, Inc. (2019)
23. Pietsch, M., et al.: A framework for multi-component analysis of diffusion MRI data over the neonatal period. Neuroimage **186**, 321–337 (2019)
24. Raffelt, D., et al.: Apparent fibre density: a novel measure for the analysis of diffusion-weighted magnetic resonance images. Neuroimage **59**(4), 3976–3994 (2012)
25. Schilling, K.G., et al.: Prevalence of white matter pathways coming into a single white matter voxel orientation: the bottleneck issue in tractography. Hum. Brain Mapp. **43**(4), 1196–1213 (2022)
26. da Silva, M.O., et al.: FOD-Swin-Net: angular super resolution of fiber orientation distribution using a transformer-based deep model. In: IEEE ISBI (2024)
27. Skare, S., et al.: Condition number as a measure of noise performance of diffusion tensor data acquisition schemes with MRI. J. Magn. Reson. **147**(2), 340–352 (2000)
28. Spears, T., Fletcher, P.T.: Learning spatially-continuous fiber orientation functions. In: IEEE ISBI (2024)
29. Tournier, J.-D., Calamante, F., Connelly, A.: MRtrix: diffusion tractography in crossing fiber regions. Int. J. Imaging Syst. Technol. **22**(1), 53–66 (2012)
30. Tournier, J.-D., Calamante, F., Connelly, A.: Robust determination of the fibre orientation distribution in diffusion MRI: non-negativity constrained super-resolved spherical deconvolution. Neuroimage **35**(4), 1459–1472 (2007)
31. Yushkevich, P.A., Gao, Y., Gerig, G.: ITK-SNAP: an interactive tool for semiautomatic segmentation of multi-modality biomedical images. In: IEEE EMBC, pp. 3342–3345 (2016)

Synthesizing 3D Axon Morphology: Springs are All We Need

Ruiqi Cui[1](✉), J. Andreas Bærentzen[1], and Tim B. Dyrby[1,2]

[1] Department of Applied Mathematics and Computer Science, Technical University of Denmark, 2800 Kongens Lyngby, Denmark
ruicu@dtu.dk
[2] Danish Research Centre for Magnetic Resonance, Centre for Functional and Diagnostic Imaging and Research, Copenhagen University Hospital - Amager and Hvidovre, 2650 Hvidovre, Denmark

Abstract. The realism of digital phantoms for the white matter microstructure is highly valued. Realistic synthesis provides reliable input to generate synthetic diffusion MRI signals for evaluating biophysical models or training machine learning models of microstructure features, such as axon diameter, shapes, and cellular structures. Inspired by the popular spring-mass systems used in physical simulation, we propose a novel and flexible method for synthesizing axon morphology and its dynamics with physical constraints. Specifically, starting with an initial axon configuration, our method constructs a spring-mass system based on specific sampling rules inspired by the real 3D axons and cell morphology observed in X-ray synchrotron imaging. By minimizing the spring potential energy, our method optimizes the positions of sampled mass points, thereby deforming the axon morphology from its physical surroundings. After the optimization, a triangle mesh of the axon surfaces is obtained and can be used as input for Monte Carlo diffusion MRI simulations. Experimental results demonstrate that our approach successfully mimics a range of axon morphologies and the dynamic environment.

Keywords: Axon morphology · Synthesis · Physics-constrained

1 Introduction

Given the fact that diffusion MRI (dMRI) is the most promising in-vivo technique for studying the micro-environment of the brain, it is crucial to accurately model the diffusion MRI signal to advance medical research. In pursuit of this goal, various biophysical models [1,9] are developed but still need to be evaluated. Concurrently, the advent of machine learning introduces new possibilities for this modeling task. Nevertheless, all models require additional data for either evaluation or training. Thus, it is of great interest to support these models by generating more reliable microstructure synthesis through numerical simulations. While the dMRI process can be simulated by Monte Carlo (MC) simulation methods [12], there is a high demand for realistic numerical phantoms of the microenvironment.

With the diffusion MRI process virtualized, a significant application is to serve as an evaluation tool for any proposed biophysical model. The digital phantom simulates the corresponding DWI signal based on data extracted from dMRI. Consequently, the accuracy of the model can be assessed by comparing both simulated and real signals.

Several existing works serve to generate the synthetic microstructure, including White Matter Generator (WMG) [15], CACTUS [14], ConFiG [4], and MEDUSA [8]. MEDUSA, which is based on sphere representation, facilitates the generation of multiple compartment types. Nevertheless, it fails to capture the eccentric shapes of axon cross sections shown by X-ray synchrotron imaging [2,10]. To allow greater eccentricity in cross sections and more realistic axon morphology modulated by the local environment [2], WMG [15] employs ellipsoids and considers dynamic aspects of the micro-environment. However, neither method accommodates non-symmetric deformation on cross-sections observed in X-ray synchrotron imaging [2]. By implementing a merging rule of spheres, ConFiG [4] enables non-symmetric and non-convex cross-sectional shapes. Meanwhile, CACTUS [14] optimizes capsule representation in parallel, enabling flexible deformation and high efficiency. Yet ConFiG [4] and CACTUS [14] only consider the interaction between axons, ignoring other influences from the microenvironment.

Despite the advances, none of these approaches consider the mechanical properties of the microenvironment, which could be one of the direct reasons for the deformation. In this work, we propose a synthesis method that supports tissue deformation based on a mechanical spring-based model. As the membranes of the white matter are expected to deform elastically, a natural choice is to use a *spring-mass system* as the physical model.

1.1 Spring-Mass System

The spring-mass system is a fundamental model in the field of physics-based simulation, widely utilized for its simplicity and ability to capture the essential dynamics of more complex physical systems. This model consists of mass points that are connected by springs, where each spring follows Hooke's Law, stating that the force exerted by the spring is proportional to its displacement from the equilibrium position. Specifically, we adopt the simplest linear spring in our design. The force F on the spring is thus calculated by $F = k \cdot |l - L|$, where k is the coefficient of elasticity, l is the length of the spring and L is the rest length of this spring. This model forms the backbone of various applications ranging from mechanical engineering to computer graphics [11]. In this study, we utilize this model to simulate various forces in the microenvironment, including surface tension (cell membrane), axon-axon or axon-extracellular-structure interactions, cytoskeletal tension, and others.

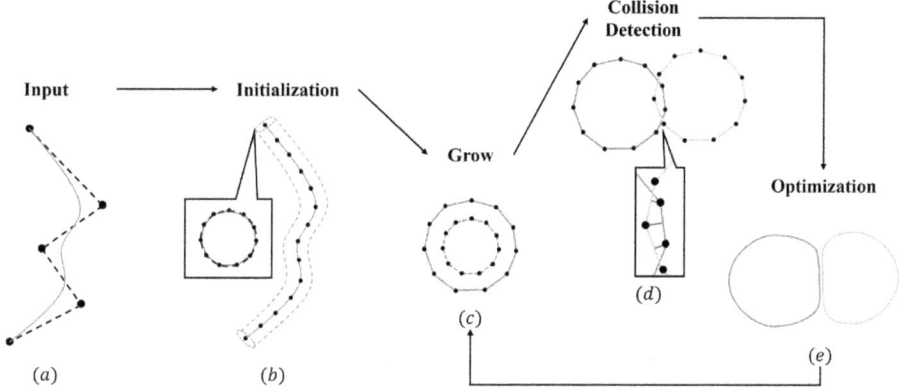

Fig. 1. Pipeline of our method. All the solid segments shown in this figure are springs. Specifically, those in red are collision springs for resolving shape intersections. Different colors represent different axons, i.e. blue and yellow. (Color figure online)

2 Method

Figure 1 illustrates the pipeline of our algorithm. Starting with a configuration file containing parameters of different non-uniform rational B-splines (NURBS) curves [6] as input, we initialize axons as curved cylinders with corresponding initial radius. Specifically, the NURBS curve [6] is a mathematical model commonly used in computer graphics and CAD to generate and represent smooth and complex curves and surfaces. At the initialization stage, according to specific sampling rules, surface and center mass points are sampled for constructing the spring-mass system. Subsequently, the axons grow along the radial direction. Following each iteration of growth, the collision detection and optimization process will deform the axons to mimic the interactions and mechanical properties by minimizing the spring energy. The source code is available upon request.

2.1 Input and Initialization

As illustrated in insets (a) and (b) of Fig. 1, the initial axons are provided as NURBS curves [6] with specific radius. Based on the initialized shape, we will construct a spring-mass system. The first step is to sample the mass points (black dots in the inset (b)) around the axon. Figure 2 gives a simplified illustration of the spring-mass system.

The center mass points are sampled first. With a pre-defined length Δx, the number of sampled points is calculated by $\frac{\sum_{i=1}^{n-1} ||P_{i+1}-P_i||}{\Delta x}$, where P_i is the coordinate of the ith control point, $||\cdot||$ is the Euclidean distance between two points. By default, in the defined voxel size of $50\,\mu\text{m} \times 50\,\mu\text{m} \times 50\,\mu\text{m}$, we use $\Delta x = 0.01\,\mu\text{m}$. We evenly sample n points along the provided NURBS curve as the center points. Each center point is associated with a ring of surface mass

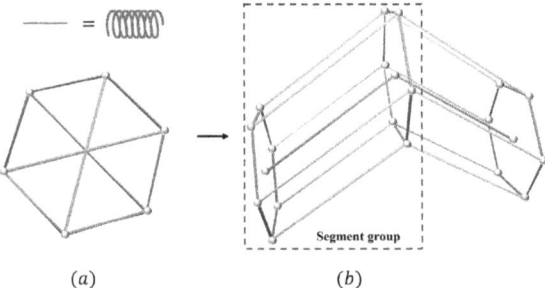

Fig. 2. Example of spring-mass system design. All the segments are springs for the spring-mass system. (a) spring connections within a ring. (b) spring connections among different rings. The spring-mass system is the combination of both insets (We rendered it separately for the clarity of presentation). Different types of springs are rendered in different colors. Green springs stand for the support force to membrane given by the cytoskeleton, blue and orange springs mimic the membrane surface tension, and red springs represent the tension of the skeleton connection. (Color figure online)

points. According to another pre-defined resolution radial interval r, the number of surface points within a ring is decided by $\frac{360}{r}$, (by default, $r = 5$).

After all mass points are sampled, we generate the springs as illustrated in Fig. 2. One can notice that the center points are disconnected from the surface points. This decision was made because we wanted to decouple the control of the axial and radial elasticity that mimics the rigidity of the cytoskeleton and cell membrane. And it greatly accelerates the convergence of the spring-mass system for both center points and surface points. Intuitively, the springs mimicking the axon's skeleton (red springs in Fig. 2) are far stiffer than those for surface tension. When two springs with significant differences in k are connected to the same mass point, the time step has to be small enough for the softer spring to take effect, which means a longer convergence process.

Furthermore, another decision is to connect only pairs of diametrically opposite points to introduce the support force cytoskeleton (green springs that connect opposite sides of the tube in Fig. 2). An ideal model would be a fully connected spring-mass system that connects all the sampled mass points. However, it would not only enlarge the parameter domain significantly but also slow the convergence.

2.2 Growth

In this synthesis, we also take into account dynamic changes, as it is believed that the morphology of the axons is affected by the dynamic local environment [2]. Thus, in each iteration, the axons grow along the radial direction according to the growth speed given by the input. The length of each radial spring (green and blue springs in Fig. 2) l is updated as follows:

$$l = l \cdot (1 + growth\_speed) \tag{1}$$

2.3 Collision Handling

After growth, the entire scene may exhibit numerous geometrical inconsistencies, such as collisions between axons. To address this, a collision handling stage is implemented to detect and resolve such issues. This stage is divided into two phases: the broad phase and the narrow phase, which is a widely adopted approach that enhances the algorithmic efficiency in computer graphics [5].

Broad Phase. The broad phase involves detecting intersections between the axis-aligned bounding boxes (AABB) of segment groups. Each segment group comprises two neighboring center points and their corresponding surface points, as illustrated by the purple dashed box in Fig. 2. By identifying all colliding AABB pairs, we can avoid performing a full collision detection for every possible segment group pair, thereby improving efficiency.

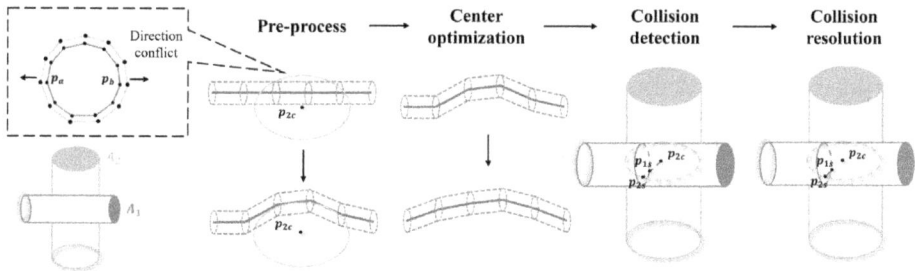

Fig. 3. Pipeline of narrow phase. A_1 and A_2 are two colliding axons shown as an example. The red solid segments represent the skeleton connection, while segments in other colors are surface connections within corresponding axons. Points belonging to A_1 are noted as p_{1x}, the same applies to points of A_2. (Color figure online)

Narrow Phase. In the narrow phase, we perform local collision detection for each colliding AABB pair identified in the broad phase. Figure 3 shows various steps involved in detecting and resolving potential collisions. First, we iteratively move the colliding center points of A_1 out of A_2. This pre-process prevents the direction conflict, depicted on the left of Fig. 3, from happening: when the center point of A_1 is inside A_2, points p_a and p_b will tend to move in opposite directions during subsequent collision resolution, which is undesirable. In each iteration, we locate the corresponding center point p_{2c} on A_2 to determine the direction of movement of the center point in a collision. This iterative design is necessary because it is not guaranteed that the colliding center point of A_1 will correspond to the same p_{2c} on A_2 after movement in 3D. Thus, we perform a minor movement and then update p_{2c}. Practically, the number of iterations is set to 50, which has been found to be sufficiently effective through experimentation.

After the pre-processing stage, we separately optimize the center spring-mass system. Without the surface points, we can employ a more relaxed line search stopping rule, i.e. Armijo condition [3]. In the final optimization, we optimize all surface springs and collision springs with the strictest stopping criterion, i.e. Strong Wolfe condition [16]. An additional precaution for the center optimization is that we constrain the moving direction of the center point within the cross-section plane perpendicular to the tangent direction. This ensures the order of sampled points remains consistent and prevents any reordering that could disrupt the integrity of the axon structure. Importantly, in the two stages described above, whenever a center point is moved, the corresponding surface points are moved in the same manner.

Subsequently, we iterate through all the surface points within each colliding AABB pair to detect if any surface point is inside the other axon. As illustrated in Fig. 3, for a surface point p_{1s} on A_1, the corresponding center point p_{2c} is identified by vector projection. Subsequently, we iterate over all surface points in the cross-sectional plane of A_2 to find the corresponding surface point p_{2s}. Since we move away from using NURBS representation after initialization in favor of a discrete setting, all surface points (p_s) are obtained through bi-linear interpolation of the sampled surface points. Similarly, all center points (p_c) are from linear interpolation of sampled center points.

Finally, if the following inequality is true, there is a collision.

$$||P_{2s} - P_{2c}|| + \text{safe\_zone} > ||P_{1s} - P_{2c}|| \qquad (2)$$

where $P_\square$ is the coordinate of the specific point, $||\cdot||$ stands for the Euclidean distance of the vectors. safe_zone is a float number defined to ensure sufficient deformation. In our work, we adaptively define the safe zone by the following equation:

$$\text{safe\_zone} = 1e^{-3} \cdot \min(r(A_1), r(A_2)) \qquad (3)$$

where $r()$ is a function to retrieve the maximum radius in the collision area of a specific axon, while A_1, A_2 represent the target axons of collision detection.

If a collision exists, we add one more spring $\{p_{1s}, p_{2s}\}$ to the spring-mass system. This spring's rest length will be set to zero, which means the p_{1s} will be pushed outside as long as the spring exists during the optimization.

2.4 Optimization

Given the prepared spring-mass system described above, we optimize it to deform the axons. The objective is to minimize the spring potential energy E_s.

$$E_s = \sum_{i=1}^{N_s} \frac{1}{2} \cdot k_i \cdot (l_i - L_i)^2 \qquad (4)$$

where N_s is the number of springs in the system, k_i is the elasticity coefficient of spring i, l_i is the length of spring i, and L_i is the rest length of spring i.

We use BFGS algorithm [7] to perform a numerical optimization. Compared with popular gradient descent algorithms, this solver offers a faster convergence and adaptive step sizes. In our setting, the parameters are the coordinate values of each sampled mass point. The gradient is easily obtained by the following equation:

$$\begin{aligned}\frac{dE_s}{dx_i} &= k_i \cdot (l_i - L_i) \cdot \frac{dl_i}{dx_i} \\ &= k_i \cdot (l_i - L_i) \cdot \frac{x_i - x_j}{\sqrt{(x_i - x_j)^2 + (y_i - y_j)^2 + (z_i - z_j)^2}} \\ &= k_i \cdot (1 - \frac{L_i}{l_i}) \cdot (x_i - x_j)\end{aligned} \quad (5)$$

where x_j is the x coordinate of the other mass point of the spring i. The same calculation applies to the y and z coordinates.

In particular, after each iteration of the optimization, we update the positions of the mass points and perform local collision detection to update the collision springs in the system. Finally, after specific post-processing, which will be introduced in Sect. 2.6, we output the synthesized scene as a triangle mesh.

2.5 Post-processing

Before it enters another iteration, two post-processing operations are performed to maintain the robustness of the system. Firstly, n-iteration Taubin smoothing [13], a smoothing method without volume shrinkage, is performed to remove high-frequency noise on the mesh ($n = 30$ in our implementation). Secondly, we update the position of the center point as the barycenter of surface points in the same ring. Afterward, we again adjust the position of the surface points to keep points on the same ring in the cross-section plane that is perpendicular to the tangent direction.

2.6 Implementation Details

Input. For the input, we use the popular JSON data format as the media of the data flow. Each axon is defined as an object with the following properties, where *dim* represents the dimension of the property and *type* for the type of data:

- *control_points* (dim: $n*3$, type: *float*): Parameter for NURBS curve. n stands for the number of mass points, the same applies to the following properties.
- *k* (dim: *Scalar*, type: *integer*): Parameter for NURBS curve.
- *weight* (dim: $n * 1$, type: *float*): Parameter for NURBS curve. Each control point has its own weight.
- *radius* (dim: $n*1$ or *Scalar*, type: *float*): The initial radius of the axon. Each control point can have a specific initial radius. When provided as a *Scalar*, the same radius applies everywhere along the axon.
- *color* (dim: $3*1$, type: *float*): The color of the axon, provided in RGB format.
- *growth_speed* (dim: *Scalar*, type: *float*): The growth speed of the axon. If not provided, it will be set by default to 0.05.

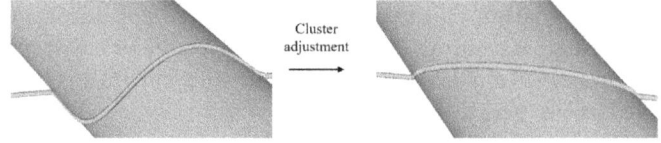

Fig. 4. Cluster adjustment to decide moving direction with global information.

Move Center - Cluster Adjustment. This pre-processing is based on per-vertex collision detection, and thus the moving direction will also be a local decision, which will introduce an artificial deformation shown in Fig. 4. To resolve this, we introduce the "cluster adjustment" to collect all neighboring colliding center points. Thus, the moving direction is decided by the vector, which is formed by the start point and end point of the cluster, and the mean corresponding center point on the other axon. This "shortest path strategy" is also aligned with what we observed from organisms in nature.

3 Results

3.1 Non-symmetric Deformation

We use a randomly generated scene containing 50 axons as an example to show the non-symmetric deformation of axons that our method provides. Due to the large difference between the implementation of different methods, we cannot compare with all the other related generators but only White Matter Generator [15] (WMG) on a similar scene. We grow the same initial axons in the scene using both methods and try to keep the axons in a similar size. In this study, we selected spring parameters characterized by high compliance and plasticity to facilitate substantial deformation.

As illustrated in Fig. 5, it turns out that the deformation of axons generated by WMG is restricted to be symmetric on the cross-sectional plane. While our method can generate more flexible deformation, a typical example can be found in the red dashed box. Notably, this experiment aims to show the versatility of our method. Under this extreme setting, it is expected to generate artificial-looking axons. Additionally, four insets are provided to illustrate more details about the generated shape by both methods.

3.2 Dynamic Synthesis with Extracellular Structure

Using the same scene, we also tested with a dynamic setting. Figure 6 shows a local detail of a dynamic scene influenced by the extracellular cells (blue solid ellipsoids). The movement (currently random) of the cells will push the axons, thus increasing the tortuosity of the axons. This feature will increase the realism since the axon morphology is modulated by not only other axons but also various

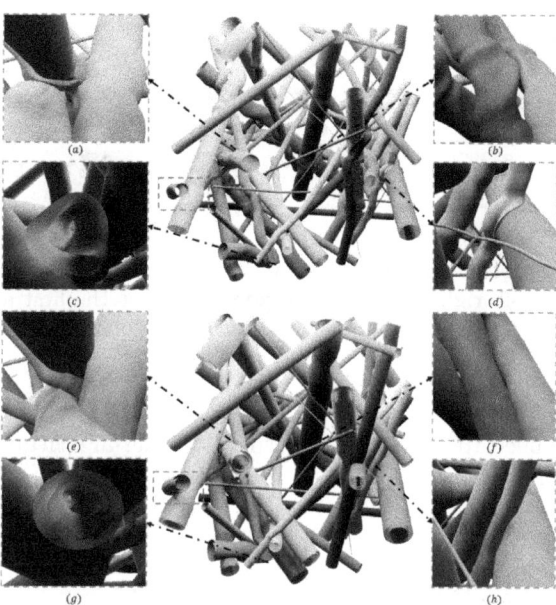

Fig. 5. A randomly generated scene with 50 axons within a voxel. The top figure is generated by our method. The bottom figure is generated by WMG [15]. Four detailed areas are provided for visual comparison. Insets (a) and (b) show local crowded areas, while (c) and (d) give examples of some non-symmetric deformation from both inside and outside view. The green axon within the red dashed box provides an example of the shortcomings of non-symmetric deformation.

Table 1. Time consumption for one iteration of growth. The experiment is performed within a voxel of $50\,\mu m \times 50\,\mu m \times 50\,\mu m$ size. The time is measured in seconds.

Radius (μm)	Number of axons				
	100	200	300	400	500
0.025	18	37	50	213	126
0.05	47	96	76	276	128
0.125	92	297	562	1766	1145

structures [2]. Currently, we only implemented cells as random ellipsoids. However, other structures can be easily integrated into our framework by sampling surface points and defining the corresponding collision function.

We tested our method on a PC with CPU 13th Gen Intel(R) Core(TM) i7-1365U, 1.80 GHz. Without any parallelism, the time spent for one iteration under different settings is illustrated in Table 1. Scenes with sparse axon distribution, such as 100/200/300 axons with a radius of $0.025\,\mu m$, achieve almost linear complexity. When heavy collisions exist, the time consumption grows significantly.

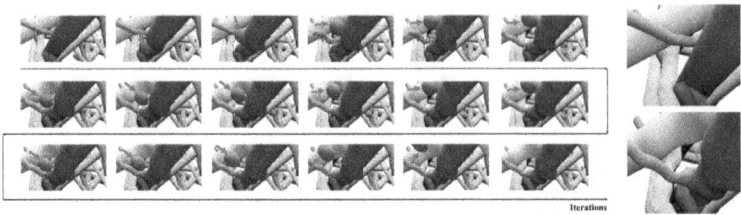

Fig. 6. A local area of a dynamic scene. In each frame, the moving cell is pointed by a red arrow. On the right, one region is zoomed in for the first and last iteration. Compared with the first iteration, the tortuosity of axons in the last iteration has significantly increased. (Color figure online)

In the future, we need to enable parallelism and other algorithmic modifications to improve the efficiency.

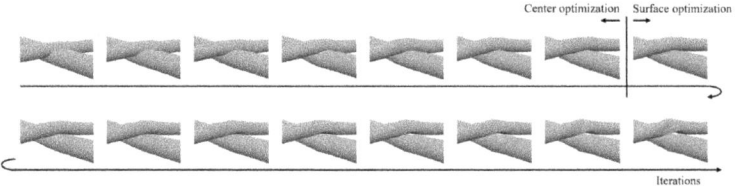

Fig. 7. The whole optimization process on a simple scene. The axon-axon interaction is mimicked (shape intersection is solved) by iteratively minimizing the spring energy.

3.3 Physics-Based Constraints

As our key contribution is to introduce a physics-based system into the synthesis process, we will illustrate an example with the whole deformation procedure of an axon. Figure 7 gives an example of a trivial scene containing two axons. Through optimization of the spring potential energy, it ends with a reasonable deformation that solves the collision.

3.4 Parameter Study

A parameter study on elasticity coefficients is performed to elaborate on the influence of different elasticity coefficient settings. By fixing the k value of surface springs, which mimics the membrane, to 2, we observe the changes in the final output while changing other k. Figure 8 exhibits different outcomes when changing k of two types of springs mimicking axial and radial rigidity. With an increasing center point springs elasticity k_c, the axon becomes more rigid in the axial direction. Similarly, a larger intracellular support spring elasticity will result in a less deformed axon in the radial direction.

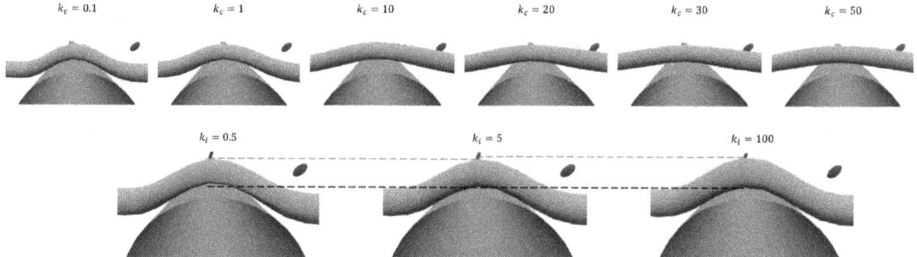

Fig. 8. Parameter study on k controlling radial and axial rigidity of the green axon. k_c is the spring coefficients of connections between skeleton points. k_i is the spring coefficient of connections inside the axon. (Color figure online)

4 Discussion

We propose a physics-based white matter synthesis method. Applying the spring-mass system, we can generate a large range of axons with different morphology by adjusting the elasticity coefficients. Compared with existing work [4,8,14,15], our method has several significant benefits. Firstly, the surface model of the spring-mass system allows non-symmetric and even extreme non-convex deformation (inset (d) in Fig. 5) within the cross-sectional plane. Secondly, our method is flexible for extension. New features/constraints can be interpreted as new springs, thus the effort for designing optimization objectives is saved.

4.1 Limitations and Future Work

Currently, most of the results are from some preliminary concept experiments. To further evaluate the performance, one key step is to involve statistics from the segmentation of the real X-ray synchrotron imaging data [2]. Moreover, one challenging process in our method is to adjust a proper ratio between different k values and the optimization time step. One tip is to fix the k of springs mimicking surface tension while tuning the other k. However, if more different springs are introduced, it will be a time-consuming process tuning them around. Moreover, the current implementation does not take care of any efficiency concerns. One improvement will be introducing parallelism into the implementation, which is straightforward based on our algorithm design.

References

1. Alexander, D.C., Dyrby, T.B., Nilsson, M., Zhang, H.: Imaging brain microstructure with diffusion MRI: practicality and applications. NMR Biomed. **32**(4), e3841 (2019)
2. Andersson, M., et al.: Axon morphology is modulated by the local environment and impacts the noninvasive investigation of its structure-function relationship. Proc. Natl. Acad. Sci. **117**(52), 33649–33659 (2020)

3. Armijo, L.: Minimization of functions having lipschitz continuous first partial derivatives. Pac. J. Math. **16**(1), 1–3 (1966)
4. Callaghan, R., Alexander, D.C., Palombo, M., Zhang, H.: Config: contextual fibre growth to generate realistic axonal packing for diffusion MRI simulation. Neuroimage **220**, 117107 (2020)
5. Ericson, C.: Real-Time Collision Detection. CRC Press, Boco Raton (2004)
6. Farin, G.: Curves and Surfaces for CAGD: A Practical Guide. Morgan Kaufmann, Burlington (2002)
7. Fletcher, R.: Practical Methods of Optimization. Wiley, New York (2000)
8. Ginsburger, K., Matuschke, F., Poupon, F., Mangin, J.F., Axer, M., Poupon, C.: Medusa: a GPU-based tool to create realistic phantoms of the brain microstructure using tiny spheres. Neuroimage **193**, 10–24 (2019)
9. Jelescu, I.O., Palombo, M., Bagnato, F., Schilling, K.G.: Challenges for biophysical modeling of microstructure. J. Neurosci. Methods **344**, 108861 (2020)
10. Lee, H.H., et al.: Along-axon diameter variation and axonal orientation dispersion revealed with 3d electron microscopy: implications for quantifying brain white matter microstructure with histology and diffusion MRI. Brain Struct. Funct. **224**, 1469–1488 (2019)
11. Nealen, A., Müller, M., Keiser, R., Boxerman, E., Carlson, M.: Physically based deformable models in computer graphics. In: Computer Graphics Forum, vol. 25, pp. 809–836. Wiley Online Library (2006)
12. Rafael-Patino, J., Romascano, D., Ramirez-Manzanares, A., Canales-Rodríguez, E.J., Girard, G., Thiran, J.P.: Robust Monte-Carlo simulations in diffusion-MRI: Effect of the substrate complexity and parameter choice on the reproducibility of results. Front. Neuroinform. **14**, 8 (2020)
13. Taubin, G.: Curve and surface smoothing without shrinkage. In: Proceedings of IEEE International Conference on Computer Vision, pp. 852–857. IEEE (1995)
14. Villarreal-Haro, J.L., et al.: Cactus: a computational framework for generating realistic white matter microstructure substrates. Front. Neuroinform. **17**, 1208073 (2023)
15. Winther, S., Peulicke, O., Andersson, M., Kjer, H.M., Bærentzen, J.A., Dyrby, T.B.: Exploring white matter dynamics and morphology through interactive numerical phantoms: the white matter generator. Front. Neuroinform. **18**, 1354708 (2024)
16. Wolfe, P.: Convergence conditions for ascent methods. SIAM Rev. **11**(2), 226–235 (1969)

Randomly COMMITting: Iterative Convex Optimization for Microstructure-Informed Tractography

Sanna Persson[1(✉)], Xinyi Wan[1,2], and Rodrigo Moreno[1,3]

[1] Department of Biomedical Engineering and Health Systems, KTH Royal Institute of Technology, Huddinge, Sweden
`sannape@kth.se`
[2] Department of Radiology and Nuclear Medicine, Erasmus MC Cancer Institute, University Medical Center, Rotterdam, The Netherlands
[3] MedTechLabs, BioClinicum, Karolinska University Hospital, Solna, Sweden

Abstract. Tractography is extensively utilized in brain connectivity studies using diffusion magnetic resonance imaging (dMRI) data. However, the presence of anatomically implausible and redundant streamlines is a significant challenge. Several tractogram filtering methods have been developed to eliminate false-positive streamlines and address these issues. This study introduces a tractography filtering method – Randomized COMMIT (rCOMMIT) – that is based on the Convex Optimization Modeling for Microstructure Informed Tractography (COMMIT) filtering method. The method aims to mitigate the biases of COMMIT for individual streamlines by assessing each streamline in multiple tractogram compositions to estimate an acceptance rate per streamline. In order to reduce the computational cost, this acceptance rate is used to create pseudo-labels that are used to train neural network classifiers in a semi-supervised manner. Specifically, we train a 1D-convolutional network on streamline characteristics, achieving an area under the receiver operating characteristic curve (AUC ROC) of approximately 90% in distinguishing between plausible and non-plausible streamlines. The results from rCOMMIT are compared with those from randomized SIFT, and the intersections between the two methods are analyzed in relation to the streamline acceptance agreement.

Keywords: Tractography · Tractogram filtering · Semi-supervised learning

1 Introduction

Diffusion magnetic resonance imaging (dMRI) has revolutionized our understanding of the brain's white matter architecture by enabling the non-invasive mapping of neural pathways through tractography. Tractography algorithms generate tractograms, which are collections of streamlines representing neural connections [3,7–9]. Applications include connectivity analysis [27,28], surgery planning [8,26] and segmentation [1,12,16,19,20,24,25]. Despite their utility, tractograms often contain anatomically implausible and redundant streamlines, posing significant challenges for accurate brain connectivity studies [2,13,19].

Many tractogram filtering methods have been developed to address these issues. These methods aim to eliminate false-positive streamlines, thereby enhancing the anatomical plausibility of the tractograms. One of the most popular is Convex Optimization Modeling for Microstructure Informed Tractography (COMMIT). COMMIT employs a global optimization approach to identify the subset of streamlines that best reconstructs the diffusion signal, effectively filtering out implausible streamlines. However, COMMIT's reliance on a global solution can introduce biases, particularly in the assessment of individual streamlines. Related tractogram filtering approaches have the same issue. Particularly, Hain et al. [6] showed that spherical-deconvolution informed tractogram filtering (SIFT) [21] can accept or reject the very same streamline depending on the composition of the tractogram.

This paper has three main contributions. First, inspired by randomized SIFT (rSIFT) [6], we introduce in this paper *randomized COMMIT* (rCOMMIT), a tractography filtering method designed to mitigate the biases inherent in COMMIT. By assessing each streamline across multiple tractogram compositions, rCOMMIT provides a more robust evaluation of streamline plausibility. This method estimates an acceptance rate for streamlines that can be used to categorize them into three groups based on their likelihood of being anatomically plausible. Second, one practical issue of both rSIFT and rCOMMIT is that they are computationally expensive. Thus, we use the acceptance rates of rSIFT and rCOMMIT as pseudo-labels to train deep learning models for streamline-wise filtering in a semi-supervised manner. Finally, we compare the results of rCOMMIT with those of rSIFT, analyzing the intersections and discrepancies between the two methods regarding streamline acceptance.

2 Method

2.1 Overview

Figure 1 shows the pipeline of rCOMMIT. The method generates multiple subsets of the original tractogram and runs the COMMIT algorithm on each subset. Each streamline's acceptance rate (AR) is calculated based on the times it is accepted across these multiple runs. We train a neural network classifier to distinguish the high-confidence streamlines from the remaining ones for a data-driven tractogram filtering.

2.2 Randomized COMMIT

The filtering result from COMMIT (as well as SIFT) for individual deadlines depends on the composition of the tractogram, both in terms of spatial distribution and the number of streamlines. In order to get the same spatial distribution as the complete tractogram, we randomly select streamlines to create multiple subsets. Then, the most important variable to consider is the subset size.

From the complete tractogram to the smallest subset, the size of subsets is chosen by approximately halving the subset size. The smallest size is set to be 2.5% of the input size since COMMIT typically filters 95–98% from tractograms with 10 million streamlines. Reducing the subset size reduces the likelihood of specific streamlines being picked. For that reason, we increase the number of subsets inversely proportional to the reduction in subset size, so every streamline is picked five times on average per subset size. The subset sizes and the number of subsets are reported in Table 1.

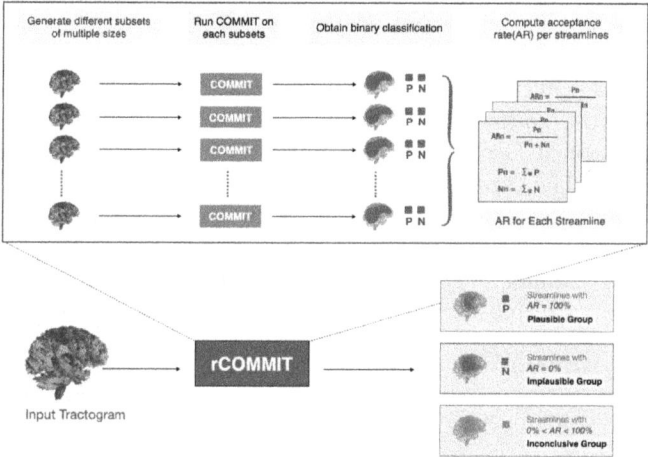

Fig. 1. Randomized COMMIT samples multiple subsets (with different sizes) of the tractogram without replacement and runs the COMMIT algorithm to filter the tractogram. With the results from filtering all subsets, the acceptance rate for each streamline is determined. The tractogram can then be divided into three groups of streamlines.

We run COMMIT with the Stick-Zeppelin-Ball model for every subset. The parameters used were: axial diffusivity of 1.7×10^{-3}, perpendicular diffusivity 0.51×10^{-3}, and isotropic diffusivities: 1.7×10^{-3} and 3×10^{-3} with tolerance 1×10^{-3} and maximum iterations 1000. Finally, the AR per streamline is computed as the ratio of the number of times it is accepted by COMMIT to the total number of assessments it undergoes.

Table 1. Subset sizes and number of subsets per subset size used for computing rSIFT and rCOMMIT.

Subset Sizes	2.5×10^4	5×10^5	6.25×10^5	1.25×10^5	2.5×10^6	5×10^6	1×10^7
Number of Subsets	200	100	80	40	20	10	5

2.3 Network Architecture

One of the issues of rCOMMIT (and rSIFT) is that it is computationally expensive. One way to reduce such a cost is by training a neural network to mimic as much as possible of rCOMMIT.

For this purpose, we used the ARs to categorize streamlines into three groups: high-confidence plausible, low-confidence plausible, and implausible with corresponding $AR = 1$, $0 < AR < 1$, and $AR = 0$. These categories serve as pseudo-labels for training machine learning models, where we train binary classifiers to distinguish the streamlines with $AR = 1$ from the other streamlines. We evaluate the classification performance on both plausible/implausible streamlines and plausible/other.

Using these pseudo-labels, we train a machine-learning model to classify streamlines based on their characteristics, such as length, curvature, and anatomical location. Following [11], we use only the coordinates of the streamlines as an input. This approach allows us to create a semi-supervised learning framework.

We use a simple 1-D convolutional network with two layers and Relu activation; see Fig. 2. The convolutional layers are followed by the dense layer with a dropout percentage of 20%. This architecture is similar to the one proposed in [6]. We normalize the streamline coordinates with the mean and standard deviation of the training subjects in each coordinate direction. The streamlines are sub-sampled with linear interpolation to the median number of points.

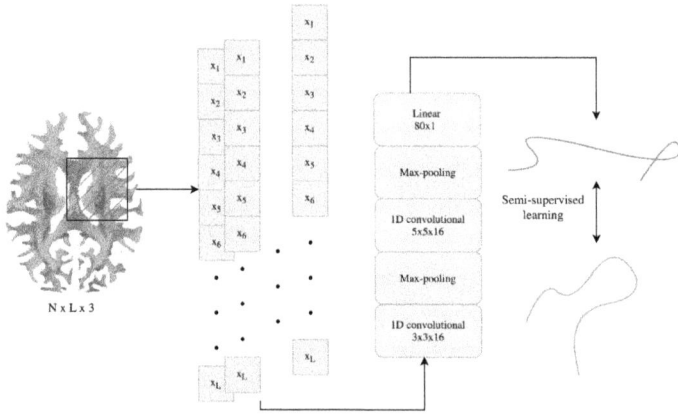

Fig. 2. Each streamline is processed as an independent 3D sequence by the model, and we obtain a prediction of its acceptance rate that is compared to the pseudo-labels.

3 Experiments

3.1 Data

We use a subset of the Human Connectome Project (HCP) from a dataset preprocessed data by Glasser et al. [5] with tractograms generated by Wasserthal et al. [25] using the iFOD2 from MRtrix3 [23]. The tractograms cover the entire white matter volume with 10 million streamlines ranging from 40 to 250 mm in length. The tractography was probabilistic with anatomical constraints using a step size of 0.625 mm. The streamlines have been compressed using the method by Presseau et al. [15] with a tolerance level of 0.35 mm.

3.2 Randomized COMMIT

We evaluate the rCOMMIT on 10 subjects from the HCP dataset. Figure 3 shows the distribution for each subject's acceptance rates. All subjects have a significant fraction of streamlines with an acceptance rate of 0 and a small fraction with an acceptance rate of 1. The data distribution is highly right-skewed, as seen in Table 2.

Table 2. Summary statistics for rCOMMIT acceptance rates. Streamlines with AR = 1 are accepted in every subset they occur in and correspondingly streamlines with AR = 0 are never accepted. The statistics are reported as percentages. Given that the data is highly right-skewed, measures of central tendency such as mean and median are not used.

Statistic	Acceptance Rate (%)
Subjects	10
25%	2.3%
50%	13%
75%	43%
% AR = 1	3.7%
% AR = 0	23.9%

3.3 Comparison with Randomized SIFT

A subset of our dataset for rCOMMIT has also been processed with rSIFT, comprising six subjects in total. We have analyzed the intersection of the results of these methods in more detail to understand their classification decisions. The rSIFT parameters are the same as in [6].

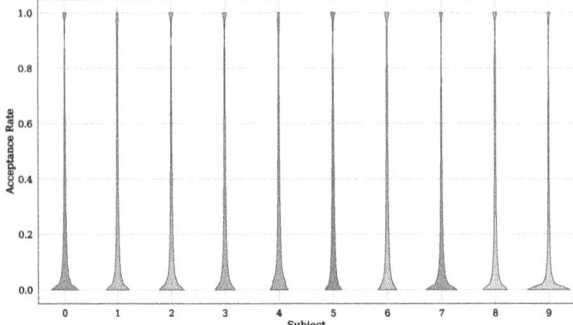

Fig. 3. The distributions of acceptance rates for each subject processed with Randomized COMMIT.

Figure 4 shows the distribution of rSIFT and rCOMMIT acceptance rates of streamlines that the other method accepts for all subjects. As shown, the two methods have a large number of streamlines where both have an acceptance rate of 1.0, but there are many other streamlines where the two methods disagree. As shown, the distributions have high concentrations around $0/1$. Further, rSIFT disregards more streamlines than rCOMMIT, and the intersection of the two sets is just 0.8%.

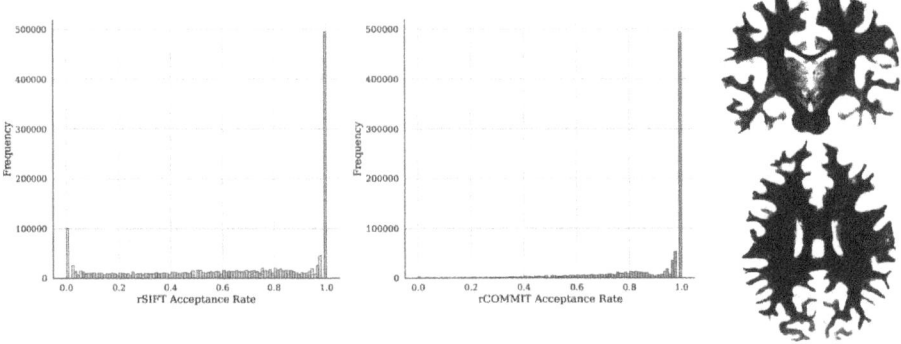

Fig. 4. Distribution of rSIFT acceptance rates (left) for accepted streamlines by rCOMMIT (acceptance rate = 1) and distribution of rCOMMIT rates for accepted streamlines by rSIFT (middle). The percentages are given with regard to the total number of streamlines from all subjects in the dataset. The tractogram (right) with the streamlines of rCOMMIT (blue) is overlayed with the intersectional streamlines (red). (Color figure online)

To compare the agreement between rSIFT and rCOMMIT, we performed an ordinary least squares regression analysis on the acceptance rates of 60 million streamlines. The dependent variable represents the acceptance rate from rSIFT, while the independent variable represents the acceptance rate from rCOMMIT.

The regression model yielded an R-squared value of 0.513, indicating that approximately 51.3% of the variance in the acceptance rates from rSIFT can be explained by the acceptance rates from rCOMMIT. The coefficient for the independent variable was 0.5706 ($p < 0.001$), indicating a statistically significant positive relationship between the acceptance rates of the two methods. The intercept of the model was -0.0103 ($p < 0.001$). However, the Omnibus and Jarque-Bera tests indicated non-normality in the residuals ($p < 0.001$), suggesting that there may be patterns in the residuals not captured by the model. We further compare the variation between rSIFT and rCOMMIT on a subject level and find a Spearman's correlation 0.62 (0.51, 0.74) and Cohen's kappa of 0.34 (0.22, 0.46), indicating moderate correlation between the methods' acceptance rates.

Comparing the results from rSIFT with rCOMMIT, we conclude that the SIFT method, in general, is more strict in filtering than COMMIT. From the results, SIFT is prone to reject more streamlines when applying both methods on the same tractogram. The difference in the filtering approaches may explain this. Both methods seek a set of streamlines that best describes the data. For COMMIT, the data to be reconstructed by the streamlines is measured diffusion data. During filtering, SIFT removes streamlines iteratively to fit the measured data better, making this process irreversible. It is also possible that the solution will reach local minima instead of global solutions.

3.4 Neural Network Models

We trained our models on three subjects and left three subjects out for testing, comprising 30 million streamlines in total. Since each streamline is processed independently and streamline properties between subjects are similar, we consider this to form a stable estimate of the model's performance. Table 3 shows the results of the two models for rSIFT and rCOMMIT.

Table 3. Results from training the model for 50 epochs on negative-vs-positive streamlines.

Model	AUC-ROC	Specificity	Sensitivity
rSIFT (P/N)	0.90	0.86	0.79
rCOMMIT (P/N)	0.80	0.71	0.73
rSIFT (P/All)	0.84	0.74	0.79
rCOMMIT (P/All)	0.73	0.61	0.73

3.5 Classifying the Intersection

To classify the intersection, we use the models trained on rSIFT and rCOMMIT labels, respectively, and fit a linear regression model on their results with our validation data consisting of one subject with 10 million streamlines. The intersection acceptance rate is then predicted as a value from the linear regression model. Positive streamlines in the intersection are when both rSIFT and rCOMMIT have $AR=1$ and negative when both have $AR=0$ with remaining streamlines being inconclusive. In Table 4, the intersection results are shown distinguishing between the different types of streamlines. We note that our models are not trained on the task of distinguishing between negative and inconclusive streamlines.

Table 4. Classification results for the intersection of rSIFT and rCOMMIT. The streamlines are divided by acceptance rates into groups: Positive for $AR=1$, Negative for $AR=0$ and Inconclusive for $0 < AR < 1$. The reported values are on 30 million streamlines for a test set consisting of three subjects.

Evaluation	AUC-ROC	Specificity	Sensitivity
Intersection (P/All)	0.85	0.71	0.82
Intersection (P/N)	0.92	0.86	0.82
Intersection (P/I)	0.82	0.77	0.72
Intersection (I/N)	0.67	0.61	0.62

3.6 Redundancy Estimation

Redundancy in tractograms is a prevalent issue [2,13,14,18], and we propose to estimate it by comparing the rCOMMIT/rSIFT and the trained classification models' predictions. The rCOMMIT and rSIFT models give acceptance rates based on sampling tractogram subsets. The neural network models instead work on a streamline-by-streamline basis. That is, some streamlines that may be disregarded in a certain tractogram composition and end up in the inconclusive group in rSIFT and rCOMMIT may be accepted when considering each streamline. In Fig. 5, the distributions of the model predictions are shown on the streamlines classified as inconclusive by rCOMMIT and rSIFT. Most streamlines are also classified by the network as inconclusive here, with a right-skewed distributions. Depending on the threshold for classifying a threshold as positive with the model, there is; however, a tail of streamlines that are predicted with higher probabilities as plausible by the model. Computing an average for thresholds for a positive streamline prediction from 0.5 to 1, we obtain redundancies of 11% (rSIFT), 13% (rCOMMIT), and 12% (intersection).

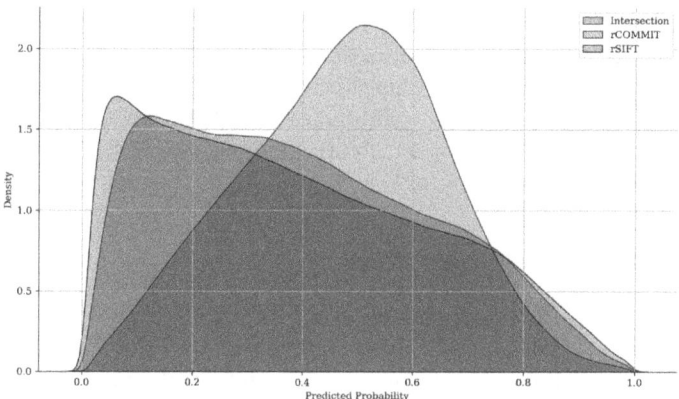

Fig. 5. Distribution of model prediction on rCOMMIT, rSIFT and intersection for streamlines classified as inconclusive. The data consists of three subjects with 30 million streamlines in total whereof 21 million were inconclusive.

4 Discussion

We introduced rCOMMIT, a filtering method to estimate a plausibility score per individual streamline. We also trained neural networks aiming at mimicking rCOMMIT while reducing the high computational cost of the method. In our comparisons of rSIFT [6] with rCOMMIT we found that rSIFT is more strict. While AR scores of the two methods overlap, the two methods yield different results. This implies that ensemble filtering schemes that combine both methods could be relevant to test in clinical applications.

We note that the networks are trained on tractograms processed with the same tractography algorithm and that more training may be required for out-of-distribution tractograms such as for deterministic tractography algorithms. Both the SIFT and COMMIT algorithms consider the composition of the tractogram in filtering. Therefore, the machine learning models are inherently limited, using only individual streamlines, and are not expected to be able to imitate the results of these algorithms completely. In this work, we use the discrepancy to estimate tractogram redundancy. We compared our classification results to a Transformer-Encoder model with three layers and found similar performance on rSIFT and rCOMMIT. This suggests that the performance limitation is related to the representation richness of the streamline coordinates rather than the model complexity. Similarly, rSIFT seems to depend more on the streamline features than rCOMMIT, considering the observed difference in classification performance in Table 3. Furthermore, there is a difference between the rCOMMIT and rSIFT prediction distributions; see Fig. 5, possibly related to the pseudo-labels signal or the observed differences between the filtering methods.

We also note that COMMIT generates a score that reflects its contribution to the diffusion signal. In rCOMMIT, this score is only used to keep streamlines with a score above zero. While using that score for rCOMMIT could be interesting, such adaptation is not straightforward. As discussed in Jörgens et al. [10], scores from COMMIT (and other methods such as SIFT2 [22] and COMMIT2 [17]) are not directly related to redundancy. It, therefore, poses the challenge of determining the boundary for which streamlines should be filtered out in the randomized approach.

4.1 Limitations and Future Work

The design of rCOMMIT presents several limitations that warrant discussion. The computation of AR depends on the number of COMMIT runs each streamline undergoes. Randomizing streamline selection could introduce bias in AR values if some streamlines are assessed fewer times. One potential solution is to ensure that only streamlines with a minimum of five assessments are included in the pseudo-ground truth. However, this may result in varying numbers of streamlines between subjects.

Future research should focus on several key areas to enhance the accuracy and applicability of the proposed methods. Firstly, incorporating additional structural features of the streamlines into the machine learning model could provide more anatomical context, potentially improving classification performance. A bundle context could be valuable to any model architecture to set each streamline into the context of the whole-brain tractogram. Additionally, investigating other tractogram filtering methods, such as SIFT2, COMMIT2, and $COMMIT_{blur}$ [4], could provide valuable insights into their similarities and differences. Similar experiments with these methods could help achieve higher-quality tractograms, facilitating more accurate brain connectivity studies. Finally, the HCP dataset can be considered high-quality. Therefore, comparing the results with those on clinical datasets would be valuable to understanding the effect of image quality on the filtering results.

5 Conclusion

In this study, we have introduced randomized COMMIT (rCOMMIT), a novel tractography filtering method designed to mitigate the biases inherent in traditional COMMIT. By assessing each streamline across multiple tractogram compositions, rCOMMIT provides a more robust evaluation of streamline plausibility. We employed a 1D-convolutional neural network model to classify streamlines based on their characteristics, achieving an area under the receiver operating characteristic curve (ROC AUC) of approximately 90%. Our results indicate that rCOMMIT offers a promising approach to improving the anatomical plausibility of tractograms. Future work should focus on incorporating additional bundle-context features and investigating other filtering methods to enhance the quality of tractograms further.

Acknowledgments. This research has been partially funded by Digital Futures, project dBrain, the Swedish Research Council, Grant No. 2022-03389, and MedTechLabs. Data were provided by the Human Connectome Project, WU-Minn Consortium (Principal Investigators: David Van Essen and Kamil Ugurbil; 1U54MH091657) funded by the 16 NIH Institutes and Centers that support the NIH Blueprint for Neuroscience Research; and by the McDonnell Center for Systems Neuroscience at Washington University.

Data and Code Availability. We used publicly available data from the Human Connectome Project and derived data analyzed in this paper can be provided upon request. Model code is available on Github.

Disclosure of Interests. The authors declare that the research was conducted in the absence of any commercial or financial relationships that could be construed as a potential conflict of interest.

References

1. Bertò, G., et al.: Classifyber, a robust streamline-based linear classifier for white matter bundle segmentation. Neuroimage **224**, 117402 (2021). https://doi.org/10.1016/j.neuroimage.2020.117402
2. Daducci, A., Dal Palú, A., Descoteaux, M., Thiran, J.P.: Microstructure informed tractography: pitfalls and open challenges. Front. Neurosci. **10** (2016). https://doi.org/10.3389/fnins.2016.00247, publisher: undefined
3. De Benedictis, A., et al.: New insights in the homotopic and heterotopic connectivity of the frontal portion of the human corpus callosum revealed by microdissection and diffusion tractography. Hum. Brain Mapp. **37**(12), 4718–4735 (2016). Publisher: Wiley Online Library
4. Gabusi, I., Battocchio, M., Bosticardo, S., Schiavi, S., Daducci, A.: Blurred streamlines: a novel representation to reduce redundancy in tractography. Med. Image Anal. **93**, 103101 (2024). https://doi.org/10.1016/j.media.2024.103101
5. Glasser, M.F., et al.: The minimal preprocessing pipelines for the Human Connectome Project. Neuroimage **80**, 105–124 (2013). https://doi.org/10.1016/j.neuroimage.2013.04.127
6. Hain, A., Jörgens, D., Moreno, R.: Randomized iterative spherical-deconvolution informed tractogram filtering. Neuroimage **278**, 120248 (2023). https://doi.org/10.1016/j.neuroimage.2023.120248
7. Hau, J., et al.: Revisiting the human uncinate fasciculus, its subcomponents and asymmetries with stem-based tractography and microdissection validation. Brain Struct. Function **222**, 1645–1662 (2017). Publisher: Springer
8. Henderson, F., Abdullah, K.G., Verma, R., Brem, S.: Tractography and the connectome in neurosurgical treatment of gliomas: the premise, the progress, and the potential. Neurosurg. Focus **48**(2), E6 (2020). Publisher: American Association of Neurological Surgeons
9. Jeurissen, B., Descoteaux, M., Mori, S., Leemans, A.: Diffusion MRI fiber tractography of the brain. NMR Biomed. **32**(4), e3785 (2019). Publisher: Wiley Online Library

10. Jörgens, D., Descoteaux, M., Moreno, R.: Challenges for tractogram filtering. In: Özarslan, E., Schultz, T., Zhang, E., Fuster, A. (eds.) Anisotropy Across Fields and Scales. MV, pp. 149–168. Springer, Cham (2021). https://doi.org/10.1007/978-3-030-56215-1_7
11. Jörgens, D., Jodoin, P.M., Descoteaux, M., Moreno, R.: Merging multiple input descriptors and supervisors in a deep neural network for tractogram filtering (2023). https://doi.org/10.48550/arXiv.2307.05786, arXiv:2307.05786 [cs]
12. Maffei, C., et al.: Using diffusion MRI data acquired with ultra-high gradient strength to improve tractography in routine-quality data. Neuroimage **245**, 118706 (2021). https://doi.org/10.1016/j.neuroimage.2021.118706
13. Maier-Hein, K.H., et al.: The challenge of mapping the human connectome based on diffusion tractography. Nat. Commun. **8**(1), 1349 (2017). https://doi.org/10.1038/s41467-017-01285-x, Publisher: Nature Publishing Group
14. Persson, S., Moreno, R.: Bounding tractogram redundancy. Front. Neurosci. **18** (2024). https://doi.org/10.3389/fnins.2024.1403804, Publisher: Frontiers
15. Presseau, C., Jodoin, P.M., Houde, J.C., Descoteaux, M.: A new compression format for fiber tracking datasets. Neuroimage **109**, 73–83 (2015). https://doi.org/10.1016/j.neuroimage.2014.12.058
16. Rheault, F., et al.: Tractostorm 2: optimizing tractography dissection reproducibility with segmentation protocol dissemination. Hum. Brain Mapp. **43**(7), 2134–2147 (2022). https://doi.org/10.1002/hbm.25777
17. Schiavi, S., et al.: A new method for accurate in vivo mapping of human brain connections using microstructural and anatomical information. Science Advances **6**(31), eaba8245 (2020). https://doi.org/10.1126/sciadv.aba8245, Publisher: American Association for the Advancement of Science
18. Schilling, K.G., et al.: Challenges in diffusion MRI tractography - lessons learned from international benchmark competitions. Magn. Reson. Imaging **57**, 194–209 (2019)
19. Schilling, K.G., et al.: Tractography dissection variability: what happens when 42 groups dissect 14 white matter bundles on the same dataset? Neuroimage **243**, 118502 (2021). https://doi.org/10.1016/j.neuroimage.2021.118502
20. Siegbahn, M., Engmér Berglin, C., Moreno, R.: Automatic segmentation of the core of the acoustic radiation in humans. Front. Neurol. **13** (2022). https://doi.org/10.3389/fneur.2022.934650, Publisher: Frontiers
21. Smith, R.E., Tournier, J.D., Calamante, F., Connelly, A.: SIFT: spherical-deconvolution informed filtering of tractograms. Neuroimage **67**, 298–312 (2013). https://doi.org/10.1016/j.neuroimage.2012.11.049
22. Smith, R.E., Tournier, J.D., Calamante, F., Connelly, A.: SIFT2: enabling dense quantitative assessment of brain white matter connectivity using streamlines tractography. Neuroimage **119**, 338–351 (2015). https://doi.org/10.1016/j.neuroimage.2015.06.092
23. Tournier, J.D., Calamante, F., Connelly, A.: Improved probabilistic streamlines tractography by 2nd order integration over fibre orientation distributions. In: Proceedings of the International Society for Magnetic Resonance in Medicine (ISMRM), vol. 1670. Wiley, Hoboken (2010). https://archive.ismrm.org/2010/1670.html
24. Warrington, S., et al.: XTRACT - standardised protocols for automated tractography in the human and macaque brain. Neuroimage **217**, 116923 (2020). https://doi.org/10.1016/j.neuroimage.2020.116923

25. Wasserthal, J., Neher, P., Maier-Hein, K.H.: TractSeg - fast and accurate white matter tract segmentation. Neuroimage **183**, 239–253 (2018). https://doi.org/10.1016/j.neuroimage.2018.07.070
26. Yang, J.Y.M., Yeh, C.H., Poupon, C., Calamante, F.: Diffusion MRI tractography for neurosurgery: the basics, current state, technical reliability and challenges. Phys. Med. Biol. **66**(15), 15TR01 (2021), Publisher: IOP Publishing
27. Yeh, C.H., Jones, D.K., Liang, X., Descoteaux, M., Connelly, A.: Mapping structural connectivity using diffusion MRI: challenges and opportunities. J. Magn. Resonan. Imaging **53**(6), 1666–1682 (2021). Publisher: Wiley Online Library
28. Zhang, F., et al.: Quantitative mapping of the brain's structural connectivity using diffusion MRI tractography: a review. Neuroimage **249**, 118870 (2022). Publisher: Elsevier

AID-DTI: Accelerating High-Fidelity Diffusion Tensor Imaging with Detail-Preserving Model-Based Deep Learning

Wenxin Fan[1,2], Jian Cheng[3], Cheng Li[1], Jing Yang[1,2], Ruoyou Wu[1,2,4], Juan Zou[5], and Shanshan Wang[1,2,4](✉)

[1] Paul C. Lauterbur Research Center for Biomedical Imaging, Shenzhen Institutes of Advanced Technology, Chinese Academy of Sciences, Shenzhen 518055, China
sophiasswang@hotmail.com
[2] University of Chinese Academy of Sciences, Beijing 100049, China
[3] Beihang University, Beijing 100191, China
[4] Peng Cheng Laboratory, Shenzhen 518055, China
[5] School of Physics and Electronic Science, Changsha University of Science and Technology, Changsha 410114, China

Abstract. Deep learning has shown great potential in accelerating diffusion tensor imaging (DTI). Nevertheless, existing methods tend to suffer from Rician noise and eddy current, leading to detail loss in reconstructing the DTI-derived parametric maps especially when sparsely sampled q-space data are used. To address this, this paper proposes a novel method, AID-DTI (**A**ccelerating h**I**gh fi**D**elity **D**iffusion **T**ensor **I**maging), to facilitate fast and accurate DTI with only six measurements. AID-DTI is equipped with a newly designed Singular Value Decomposition-based regularizer, which can effectively capture fine details while suppressing noise during network training by exploiting the correlation across DTI-derived parameters. Additionally, we introduce a Nesterov-based adaptive learning algorithm that optimizes the regularization parameter dynamically to enhance the performance. AID-DTI is an extendable framework capable of incorporating flexible network architecture. Experimental results on Human Connectome Project (HCP) data consistently demonstrate that the proposed method estimates DTI parameter maps with fine-grained details and outperforms other state-of-the-art methods both quantitatively and qualitatively.

Keywords: Diffusion tensor imaging · deep learning · SVD

1 Introduction

Diffusion magnetic resonance imaging (dMRI) is a prominent non-invasive neuroimaging technique for measuring tissue microstructure. Among various dMRI techniques, diffusion tensor imaging (DTI) [2] is widely used to extract brain tissue properties and identify white matter tracts in vivo. The metrics from DTI, such as fractional anisotropy (FA), mean diffusivity (MD), and axial diffusivity (AD) [12] have great specificity in mapping the microstructural changes caused by normal aging [30], neurodegeneration [37], and psychiatric disorders [45].

To increase the accuracy of DTI-derived parametric maps, studies typically need more than the minimum of 6 diffusion weighting (DW) directions or acquire repeated observations of the same set of DW directions [21]. Moreover, the low signal-to-noise ratio (SNR) poses significant challenges to subsequent analysis, which further increases the demand for data to enable high-fidelity DTI metrics. Therefore, there is an urgent need to develop high-quality DTI metrics estimation from sparsely sampled q-space data.

Recently, deep learning has emerged as a powerful tool for accelerating DTI imaging. The pioneering work, q-space deep learning (q-DL) [19], was introduced to directly map a subset of diffusion signals to Diffusion Kurtosis Imaging (DKI) parameters using a three-layer multilayer perceptron (MLP). Gibbons et al. [18] used a 2D convolutional neural network (CNN) to estimate the Neurite Orientation Dispersion and Density Imaging (NODDI) and generalized fractional anisotropy maps. Similarly, SuperDTI [22] used deep CNN to model the nonlinear relationship between the acquired DWIs and the desired DTI-derived maps. In addition to data-driven mapping approaches, there has been a growing interest in model-driven neural networks that leverage domain knowledge to enhance network performance and interpretability. A notable example is the works proposed by Ye et al. [41,42,44] which unfold the iterative optimization process for parameter mappings. Chen et al. used a subset q-space to estimate the parameters by explicitly considering the q-space geometric structure with a graph neural network (GNN) [6,7]. Furthermore, some excellent works in DWI super-angular-resolution can assist in the prediction of high-quality DTI metrics [8,35,38].

Despite the progress made, the current methods still suffer from noise corruption or fine detail loss at a highly accelerated imaging rate. In this study, we propose a novel model-based deep learning method, named AID-DTI (**A**ccelerating h**I**gh fi**D**elity **D**iffusion **T**ensor **I**maging) to facilitate fast and accurate DTI. The main contributions of this work can be summarized as follows:

1. We propose a simple but effective model-based deep learning model, with a newly designed regularization to facilitate high-fidelity DTI metrics derivation. This term leverages the correlations and data redundancy between metrics, specifically targeting the alignment between predicted parameters and ground truth in singular-value subspaces, thus effectively capturing fine-grained details while suppressing noise.
2. We propose a novel Nesterov-based hyperparameter adaptive learning algorithm that integrates approximate second-order derivative information into the network training process, enabling more efficient hyperparameter tuning and better performance.
3. AID-DTI enables fast and high-fidelity DTI metrics estimation using a minimum of six measurements along uniform diffusion-encoding directions. Experiments demonstrated that our method outperforms current state-of-the-art methods both quantitatively and qualitatively.

2 Methods

In this section, we present the statement of the problem and a detailed presentation of the proposed AID-DTI, which investigates the accurate DTI-derived metric estimation using only six measurements instead of the recommended 30 measurements [20] to achieve reliable prediction within the needed clinical accuracy. The proposed method is depicted in Fig. 1 and encompasses key elements, specifically the SVD-based regularization (SVD-Reg) and the Nesterov-based adaptive learning algorithm (NALA).

As illustrated in Fig. 1, the overall architecture consists of two branches, with the upper branch representing the ground truth acquisition from dense sampling, while the lower branch symbolizes the network prediction from the sparse sampling. The network input is super sparse measurements uniformly sampled from the dense measurements using DMRITool [10,11], and Singular Value Decomposition (SVD) is applied to both network prediction and ground truth to ensure the singular value consistency.

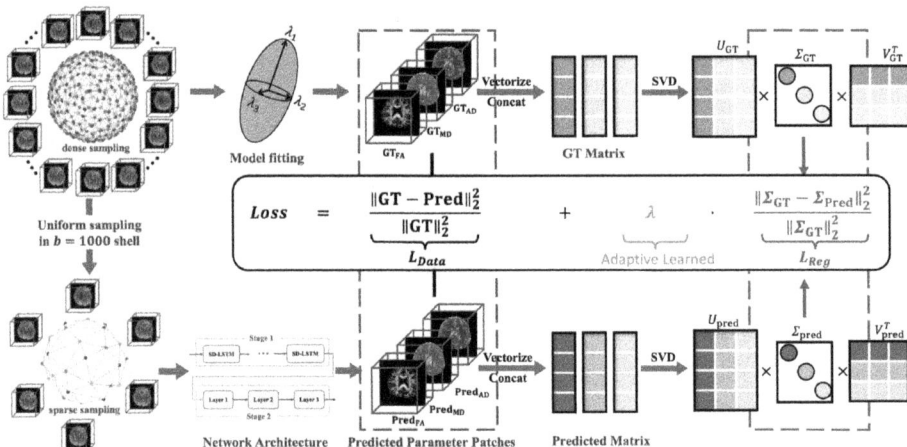

Fig. 1. The proposed AID-DTI pipeline. The network input is super sparse measurements uniformly sampled from the dense measurements, and then the mapping between the sparsely sampled signal and three DTI metrics is directly learned simultaneously. After the network output, we vectorize each parameter and concatenate them into a new matrix, then perform SVD on this matrix to obtain the singular values. The weighted parameter λ is adaptively learned to balance between data fidelity and SVD-regularization.

2.1 Task Formulation

Our goal is to estimate reliable and fine-grained DTI parametric maps using only six measurements. Each diffusion signal can be considered as a set of $W \times H \times S$ size volumes captured in the q-space. Thus the dMRI data are 4D signals of size

$\mathbb{R}^{W \times H \times S \times D}$, where W, H, S, D refer to the width, height, slice, and gradient directions, respectively.

Given the diffusion MRI data $X \in \mathbb{R}^{W \times H \times S \times D_{Full}}$ containing the full measurements in the q-space, the ground-truth scalar maps Y_{GT} obtained from all the diffusion data, we aim to design a network $\mathcal{F}_\theta$ parameterized by θ to learn a mapping from the given sparsely sampled signal $\tilde{X} \in \mathbb{R}^{W \times H \times S \times D_{Sparse}}$ to predicted DTI metrics Y, s.t $Y = \mathcal{F}_\theta(\tilde{X}) \to Y_{GT}$.

2.2 SVD-Based Regularization

Most existing regularization strategies only consider the properties of the diffusion signal and apply the regularization to the DWI data rather than the desired parametric maps, such as sparsity [9,31,32,40], low-rank [9,28,33,43], total variation [23,33,36] regularization, etc. To facilitate accurate and fine-grained DTI metric prediction, we explicitly consider the quality of derived parameters and propose the incorporation of an SVD-based regularization term to enhance performance.

$$Loss = L_{Data} + \lambda \cdot R = \frac{\|Y_{GT} - Y_{Pred}\|_2^2}{\|Y_{GT}\|_2^2} + \lambda \cdot \frac{\|\Sigma_{GT} - \Sigma_{Pred}\|_2^2}{\|\Sigma_{GT}\|_2^2} \quad (1)$$

The actual input in our implementation is the $N \times N \times N$ patches instead of the whole DWI volume, so the output of the network Y_{Pred} is $N \times N \times N \times 3$, the last dimension indicating three parameter maps. Then, we vectorize each parameter and concatenate them into a new matrix, referred to as the parameter matrix. We perform SVD on the predicted matrix and GT matrix respectively to obtain the singular values Σ_{Pred} and Σ_{GT}.

From the statistical point of view, the singular matrices of a data matrix represent the principal component directions, i.e., the directions that exhibit the highest variance corresponding to the largest singular values. According to the Eckart-Young theorem [15], the dominant singular subspaces capture the majority of the informational content. It can be believed that the major singular values encapsulate the dominant features of the three parameters. Therefore, ensuring the consistency of the primary singular values preserves the integrity of the extracted significant information, effectively maintaining fine details while reducing a certain level of noise. Subsequent denoising experiments also demonstrated the superiority of the proposed method in noise handling.

2.3 Nesterov-Based Hyperparameter Adaptive Learning Algorithm

The total loss is the weighted combination of the data-fidelity term and the proposed regularization term. However, the process of hyperparameter selection is in practice often based on trial-and-error and grid or random search [4,14,17,24], which can be a time-consuming process.

Building upon the foundation laid by previous studies [1,3,5,29], we propose a Nesterov-based hyperparameter adaption algorithm. The hyperparameter optimization problem is inherently a bilevel optimization task because of its hierarchical nature [16,17,34]. The outer problem requires minimizing the validation set loss, concerning the hyperparameter λ, and the inner problem requires minimizing the training set loss, for the model parameter θ. Thus, our method optimizes the network parameter θ and hyperparameter λ alternately on the training and validation sets, respectively, which means the λ that minimizes the validation loss will be accepted. Let λ_t and θ_t be the values of λ and θ at the step t. More specifically, the iterations go as follows:

$$\begin{cases} L(\theta, \lambda) = L_{Data}(\theta) + \lambda \cdot R(\theta) \\ \text{On Training Set: } \theta_{t+1} = \arg\min_\theta L(\theta, \lambda_t) \\ \text{On Validation Set: } \lambda_{t+1} = \arg\min_\lambda L(\theta_{t+1}, \lambda) \end{cases} \quad (2)$$

In analogy to updating network parameter θ, λ should be updated in the direction of the gradient of the $Loss\,(\theta, \lambda)$ concerning λ, scaled by another hyper-hyperparameter β. One way to compute $\frac{\partial L(\theta_{t+1}, \lambda_t)}{\partial \lambda_t}$ is the direct manual computation of the partial derivative:

$$\beta \cdot \frac{\partial Loss\,(\theta_{t+1}, \lambda)}{\partial \lambda} = \beta \cdot \frac{\partial \left[L_{Data}\,(\theta_{t+1}) + \lambda \cdot R\,(\theta_{t+1})\right]}{\partial \lambda} = \beta \cdot R(\theta_{t+1}) \quad (3)$$

In other words, the adjustment at step $t+1$ depends on the regularization term value. This expression lends itself to a simple and efficient implementation: simply remember the past regularization value. By leveraging insights from the Nesterov accelerated gradient (NAG) [26,27], which has a provably bound for convex, non-stochastic objectives, we introduce an improved momentum term m here:

$$\begin{cases} m_{t+1} = \beta \cdot m_t + R(\theta_{t+1}) + \beta \cdot [R(\theta_{t+1}) - R(\theta_t)] \\ \lambda_{t+1} = \lambda_t - \kappa \cdot m_{t+1} \end{cases} \quad (4)$$

where $R(\theta_{t+1}) - R(\theta_t)$ is actually the differential of the gradient concerning λ, which approximates the second-order derivative of the objective function. Thus, the improved momentum term m_{t+1} is the combination of the past search directions m_t, current stochastic gradient $R(\theta_{t+1})$, and the approximate second-order derivative $R(\theta_{t+1}) - R(\theta_t)$.

2.4 Backbone Network

Here, the Microstructure Estimation with Sparse Coding using Separable Dictionary (MESC-SD) [42], an unfolding network based on sparse LSTM units [46] with two cascaded stages, is employed. The first stage computes the spatial-angular sparse representation of the diffusion signal while the second stage maps the sparse representation to tissue microstructure estimates. Note that AID-DTI is highly versatile, such that networks producing output in a matrix or higher-dimensional tensor form are compatible with our methodology. We can support CNN [18], MESC-SD [42], and even one-dimensional networks like q-DL [19] can benefit from our method by appropriately reshaping their outputs into matrix form.

3 Experiments and Results

3.1 Dataset

Pre-processed whole-brain diffusion MRI data from the publicly available Human Connectome Project (HCP) dataset were used for this study [39]. Our dataset consists of 111 subjects randomly selected from the HCP, which is partitioned into 60 subjects for training, 17 subjects for validation, and 34 subjects for testing.

To obtain the input data of AID-DTI, DWI volumes acquired along six uniform diffusion-encoding directions at $b = 1000\,\text{s/mm}^2$ of each subject were selected using DMRITool [10,11]. To obtain the ground-truth DTI metrics, diffusion tensor fitting was performed on all the diffusion data using ordinary linear squares fitting implemented in the DIPY[1] software package to derive the FA, MD, and AD [12].

3.2 Comparison Methods

Our method was compared qualitatively and quantitatively with DIPY, which represents the conventional DTI model fitting (MF) algorithm, and deep learning-based approaches, including the q-DL in Golkov et al. [14], CNN in Gibbons et al. [15], MESC-SD in Ye et al. [19].

3.3 Implementation Details

The neural network was implemented using the PyTorch library (codes will be available online upon acceptance of the paper). We trained the network with Two Tesla V100 GPUs (NVIDIA, Santa Clara, CA) with 32 GB memory. All networks adopted the Adam optimizer and the learning rates were initialized as 0.01, 0.001, and 0.0001 respectively. For all the networks, the extracted brain masks from the preprocessing pipeline were applied to only include voxels within the brain when evaluating the performance.

3.4 Evaluation Results

We evaluate the performance of AID-DTI through a comparative analysis with baseline methods and other state-of-the-art methods by computing the SSIM and PSNR to quantify the similarity compared to the ground truth. The experimental results are summarized in Table 1, where it can be observed that AID-DTI surpasses the comparison methods by a large margin.

---

[1] https://github.com/dipy.

Table 1. The quantitative results were obtained with 6 diffusion directions at b-values of $1000\,\text{s/mm}^2$ in terms of MSE, SSIM, and PSNR. The best results are in **bold**.

Methods	MSE $\times 10^{-3}$				SSIM				PSNR			
	FA	MD	AD	All	FA	MD	AD	All	FA	MD	AD	All
MF	36.00	15.55	26.38	25.98 ± 3.74	0.719	0.760	0.714	0.772 ± 0.031	14.437	18.082	15.786	15.854 ± 0.618
q-DL	2.293	1.160	1.419	1.623 ± 0.23	0.904	0.952	0.931	0.929 ± 0.009	26.397	29.354	28.481	27.894 ± **0.591**
CNN	1.184	0.726	0.934	0.948 ± 0.16	0.941	0.968	0.951	0.953 ± 0.007	29.266	31.390	30.299	30.232 ± 0.679
MESC-SD	0.756	0.671	0.824	0.750 ± 0.13	0.952	0.971	0.958	0.960 ± 0.006	31.216	31.733	30.841	31.248 ± 0.689
Ours	**0.683**	**0.626**	**0.774**	**0.694 ± 0.12**	**0.956**	**0.973**	**0.961**	**0.963 ± 0.005**	**31.653**	**32.037**	**31.113**	**31.638** ± 0.676

For qualitative analysis, we provide the estimation results in Fig. 2. As can be seen, the conventional method MF produces significant estimation error and loses anatomical information when only six measurements were employed. The results from the figure also show that the q-DL method yields a relatively low signal-to-noise ratio, while the CNN method, although achieving better results, appears overly smooth in qualitative images, leading to the loss of texture. MESC-SD, as one of the state-of-the-art microstructure estimation methods, showcases excellent results, when our method combined with it, demonstrates enhanced performance as evidenced by the error maps, effectively preserving crucial anatomical details (Fig. 3).

To validate the noise-handling capabilities of the proposed method, we synthesize noisy data by introducing Rician noise at levels of 2.5% and 5% into the diffusion-weighted signals. Then input the noise-corrupted data into the trained networks to predict the three DTI-derived scalar maps. Table 2 shows the comparative results for varying noise levels. To ensure a fair comparison, we also considered applying denoising algorithms after MF, specifically BM4D[2] [13,25] was chosen in our experiments. Both Table 1 and Table 2 demonstrate that our method outperforms others in clean and noisy conditions, indicating a degree of noise resistance.

Table 2. Quantitative evaluation of denoising performance using synthetic data with different level of Rician noise. The best results are in **bold**.

Methods	$\sigma = 0.025$			$\sigma = 0.05$		
	MSE $\times 10^{-3}$	SSIM	PSNR	MSE $\times 10^{-3}$	SSIM	PSNR
MF	28.32 ± 3.57	0.760 ± 0.028	15.511 ± **0.538**	28.67 ± 3.58	0.759 ± 0.029	15.425 ± **0.540**
MF+BM4D	28.07 ± 3.56	0.761 ± 0.029	15.551 ± 0.542	28.59 ± 3.58	0.759 ± 0.029	15.437 ± 0.542
q-DL	1.604 ± 0.27	0.914 ± 0.012	27.290 ± 0.600	2.560 ± 0.38	0.882 ± 0.016	25.858 ± 0.619
CNN	1.111 ± 0.17	0.942 ± 0.008	29.545 ± 0.652	1.609 ± **0.25**	0.917 ± **0.011**	27.934 ± 0.663
MESC-SD	0.941 ± 0.16	0.947 ± 0.008	30.266 ± 0.682	1.586 ± **0.25**	0.922 ± 0.012	28.000 ± 0.655
Ours	**0.893 ± 0.15**	**0.952 ± 0.007**	**30.547** ± 0.680	**1.499** ± 0.27	**0.927** ± 0.012	**28.302** ± 0.732

---

[2] https://pypi.org/project/bm4d/.

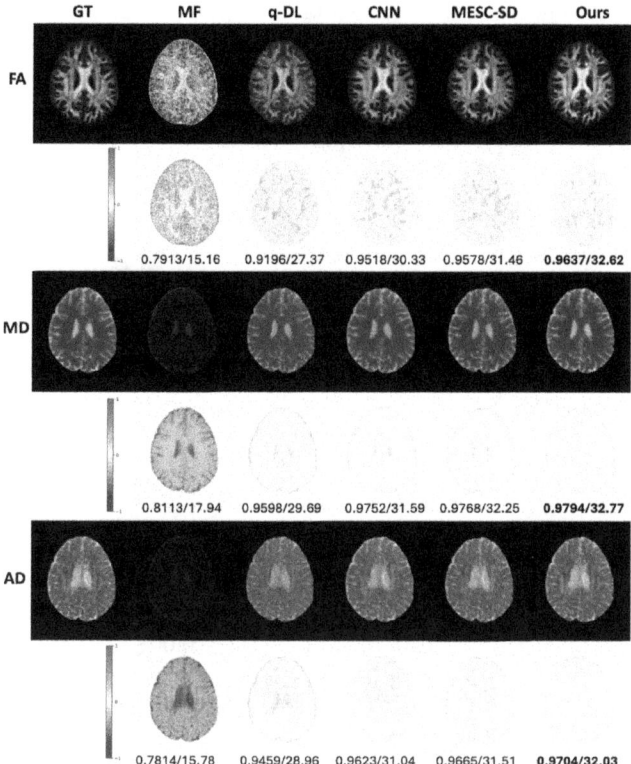

Fig. 2. The ground truth, estimated DTI parameters FA, AD, and MD, and corresponding residual maps based on MF, q-DL, CNN, MESC-SD (baseline), and Ours in a test subject with 6 diffusion directions at b-values of $1000\,\text{s/mm}^2$.

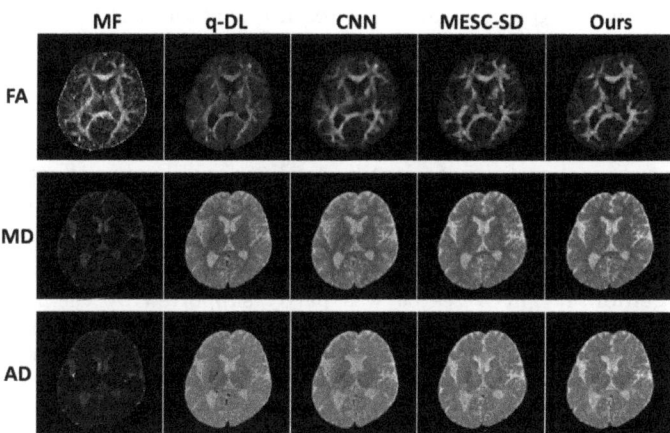

Fig. 3. Prospective results in a test subject with real low angular resolution data (6 diffusion directions at b-values of $1000\,\text{s/mm}^2$ and 2 at b_0).

Due to the absence of publicly available super low angular resolution (six-direction) datasets, we used in-house data here to conduct prospective experiments. The imaging protocol was as follows: 2 b0 gradient directions and 6 b = 1000 s/mm² gradient directions; 140 × 140 imaging matrix; voxel size 1.5 × 1.5 × 1.0 mm³; TE/TR = 66.0/5,820 ms.

3.5 Ablation Study

In this section, we perform an extensive ablation study to investigate the effectiveness of the SVD-based regularization (SVD-Reg) module and Nesterov-based adaptive learning algorithm (NALA). As shown in Table 3, the ablation study is completed under the condition of 6 gradients at a b-value of 1000 s/mm². Table 2 shows the quantitative results of the three variants, respectively. According to the quantitative results, the average values of PSNR and SSIM achieved by AID-DTI are the highest among the three variants.

Table 3. Ablation results using MSE, SSIM, and PSNR. The best results are in **bold**.

Models	SVD-Reg	NALA	MSE ($\times 10^{-3}$)	SSIM	PSNR
(A)			0.750 ± 0.13	0.960 ± 0.006	31.248 ± 0.689
(B)	✓		0.704 ± 0.12	0.962 ± 0.005	31.525 ± 0.690
(C)	✓	✓	**0.694 ± 0.12**	**0.963 ± 0.005**	**31.638 ± 0.676**

4 Conclusion

In this study, we develop a novel model-driven deep learning approach AID-DTI for reducing the q-space sampling requirement of DTI. Our method maps one b=0 image and six DWI volumes to high-quality DTI metrics employing an SVD-based regularization and introduces an adaptive algorithm for automatically updating regularization parameters. The proposed method exhibits simplicity, flexibility and has a high potential to become a practical tool in a wide range of clinical and neuroscientific applications. Future efforts will expand the proposed method to other diffusion models and more multi-parametric MR imaging scenarios.

Acknowledgements. This research was partly supported by the National Natural Science Foundation of China (62222118, U22A2040), Guangdong Provincial Key Laboratory of Artificial Intelligence in Medical Image Analysis and Application (2022B1212010011), Shenzhen Science and Technology Program (RCYX20210706092104034, JCYJ20220531100213029), and Key Laboratory for Magnetic Resonance and Multimodality Imaging of Guangdong Province (2023B1212060052).

References

1. Almeida, L.B., et al.: Parameter adaptation in stochastic optimization. In: Online Learning in Neural Networks, pp. 111–134 (1999)
2. Basser, P.J., Mattiello, J., LeBihan, D.: MR diffusion tensor spectroscopy and imaging. Biophys. J. **66**(1), 259–267 (1994)
3. Baydin, A.G., et al.: Online learning rate adaptation with hypergradient descent. arXiv preprint arXiv:1703.04782 (2017)
4. Bengio, Y.: Gradient-based optimization of hyperparameters. Neural Comput. **12**(8), pp. 1889–1900 (2000)
5. Chandra, K., et al.: Gradient descent: the ultimate optimizer. In: Advances in Neural Information Processing Systems, vol. 35, pp. 8214–8225 (2022)
6. Chen, G., G., et al.: Estimating tissue microstructure with undersampled diffusion data via graph convolutional neural networks. In: International Conference on Medical Image Computing and Computer-Assisted Intervention, pp. 280–290. Springer, Cham (2020)
7. Chen, G., et al.: Hybrid graph transformer for tissue microstructure estimation with undersampled diffusion MRI data. In: International Conference on Medical Image Computing and Computer-Assisted Intervention, pp. 113–122. Springer, Cham (2022)
8. Chen, Z., et al.: Super-resolved q-space learning of diffusion MRI. Med. Phys. (2023)
9. Cheng, J., et al.: Joint 6D k-q space compressed sensing for accelerated high angular resolution diffusion MRI. In: International Conference on Information Processing in Medical Imaging, pp. 782–793. Springer, Cham (2015)
10. Cheng, J., et al.: Novel single and multiple shell uniform sampling schemes for diffusion MRI using spherical codes. In: MICCAI 2015, Part I 18, pp. 28–36. Springer, Cham (2015)
11. Cheng, J., et al.: Single-and multiple-shell uniform sampling schemes for diffusion MRI using spherical codes. IEEE Trans. Med. Imaging **37**(1), 185–199 (2017)
12. Curran, K.M., Emsell, L., Leemans, A.: Quantitative DTI measures. In: Diffusion Tensor Imaging: A Practical Handbook, pp. 65–87 (2016)
13. Dabov, K., et al.: Image denoising by sparse 3-D transform-domain collaborative filtering. IEEE Trans. Image Process. **16**(8), 2080–2095 (2007)
14. Duchi, J., Hazan, E., Singer, Y.: Adaptive subgradient methods for online learning and stochastic optimization. J. Mach. Learn. Res. **12**(7) (2011)
15. Eckart, C., Young, G.: The approximation of one matrix by another of lower rank. Psychometrika **1**(3), 211–218 (1936)
16. Franceschi, L., et al.: Bilevel programming for hyperparameter optimization and meta-learning. In: International Conference on Machine Learning, pp. 1568–1577. PMLR (2018)
17. Franceschi, L., et al.: Forward and reverse gradient-based hyperparameter optimization. In: International Conference on Machine Learning, pp. 1165–1173. PMLR (2017)
18. Gibbons, E.K., et al.: Simultaneous NODDI and GFA parameter map generation from subsampled q-space imaging using deep learning. Magn. Resonan Med. **81**(4), 2399–2411 (2019)
19. Golkov, V., et al.: Q-space deep learning: twelve-fold shorter and model-free diffusion MRI scans. IEEE Trans. Med. Imaging **35**(5), 1344–1351 (2016)

20. Jones, D.K., Knösche, T.R., Turner, R.: White matter integrity, fiber count, and other fallacies: the do's and don'ts of diffusion MRI. Neuroimage **73**, 239–254 (2013)
21. Landman, B.A., et al.: Effects of diffusion weighting schemes on the reproducibility of DTI-derived fractional anisotropy, mean diffusivity, and principal eigenvector measurements at 1.5 T. Neuroimage **36**(4), 1123–1138 (2007)
22. Li, H., et al.: SuperDTI: ultrafast DTI and fiber tractography with deep learning. Magn. Resonan. Med. **86**(6), 3334–3347 (2021)
23. Liu, R.W., et al.: Generalized total variation-based MRI Rician denoising model with spatially adaptive regularization parameters. Magn. Resonan. Imaging **32**(6), 702–720 (2014)
24. Maclaurin, D., Duvenaud, D., Adams, R.: Gradient-based hyperparameter optimization through reversible learning. In: International Conference on Machine Learning, pp. 2113–2122. PMLR (2015)
25. Maggioni, M., et al.: Nonlocal transform-domain filter for volumetric data denoising and reconstruction. IEEE Trans. Image Process. **22**(1), 119–133 (2012)
26. Nesterov, Y.: Gradient methods for minimizing composite functions. Math. Program. **140**(1), 125–161 (2013)
27. Nesterov, Y.E.: A method of solving a convex programming problem with convergence rate $\bigl(k^2\bigr)$. Doklady Akademii Nauk 269(3), 543–547. Russian Academy of Sciences (1983)
28. Ramos-Llordén, G., et al.: SNR-enhanced diffusion MRI with structure preserving low-rank denoising in reproducing kernel Hilbert spaces. Magn. Resonan. Med. **86**(3), 1614–1632 (2021)
29. Rubio, D.M.: Convergence analysis of an adaptive method of gradient descent. M.Sc. thesis, University of Oxford, Oxford (2017)
30. Salat, D.H., et al.: Age-related changes in prefrontal white matter measured by diffusion tensor imaging. Ann. New York Acad. Sci. **1064**(1), 37–49 (2005)
31. Schwab, E., Vidal, R., Charon, N.: Joint spatial-angular sparse coding for dMRI with separable dictionaries. Med. Image Anal. **48**, 25–42 (2018)
32. Schwab, E., et al.: Global optimality in separable dictionary learning with applications to the analysis of diffusion MRI. SIAM J. Imaging Sci. **12**(4), 1967–2008 (2019)
33. Shi, F., et al.: Super-resolution reconstruction of diffusion-weighted images using 4D low-rank and total variation. In: MICCAI 2015, pp. 15–25. Springer, Cham (2016)
34. Sinha, A., Malo, P., Deb, K.: A review on bilevel optimization: from classical to evolutionary approaches and applications. IEEE Trans. Evol. Comput. **22**(2), 276–295 (2017)
35. Tang, Z., et al.: High angular diffusion tensor imaging estimation from minimal evenly distributed diffusion gradient directions. Front. Radiol. **3** (2023)
36. Teh, I., et al.: Improved compressed sensing and super-resolution of cardiac diffusion MRI with structure-guided total variation. Magn. Resonan. Med. **84**(4), 1868–1880 (2020)
37. Thompson, P., et al.: Effectiveness of regional DTI measures in distinguishing Alzheimers disease, MCI, and normal aging (2013)
38. Tian, Q., et al.: DeepDTI: high-fidelity six-direction diffusion tensor imaging using deep learning. NeuroImage **219**, 117017 (2020)
39. Van Essen, D.C., et al.: The WU-Minn human connectome project: an overview. Neuroimage **80**, 62–79 (2013)

40. Yap, P.-T., Zhang, Y., Shen, D.: Multi-tissue decomposition of diffusion MRI signals via l_0 sparse-group estimation. IEEE Trans. Image Process. **25**(9), 4340–4353 (2016)
41. Ye, C., Li, X., Chen, J.: A deep network for tissue microstructure estimation using modified LSTM units. Med. Image Anal. **55**, 49–64 (2019)
42. Ye, C., Li, Y., Zeng, X.: An improved deep network for tissue microstructure estimation with uncertainty quantification. Med. Image Anal. **61**, 101650 (2020)
43. Zhang, C., et al.: Acceleration of three-dimensional diffusion magnetic resonance imaging using a kernel low-rank compressed sensing method. Neuroimage **210**, 116584 (2020)
44. Zheng, T., et al.: A microstructure estimation Transformer inspired by sparse representation for diffusion MRI. Med. Image Anal. **86**, 102788 (2023)
45. Zheng, Z., et al.: DTI correlates of distinct cognitive impairments in Parkinson's disease. Hum. Brain Mapp. **35**(4), 1325–1333 (2014)
46. Zhou, J.T., et al.: Sc2net: sparse LSTMs for sparse coding. In: Proceedings of the AAAI Conference on Artificial Intelligence, vol. 32 (2018)

Multi-dimensional Parameter Space Exploration for Streamline-Specific Tractography

Ruben Vink[✉], Anna Vilanova, and Maxime Chamberland

Department of Computer Science and Mathematics, Eindhoven University of Technology, Eindhoven, The Netherlands
r.vink@tue.nl

Abstract. One of the unspoken challenges of tractography is choosing the right parameters for a given dataset or bundle. In order to tackle this challenge, we explore the multi-dimensional parameter space of tractography using streamline-specific parameters (SSP). We 1) validate a state-of-the-art probabilistic tracking method using per-streamline parameters on synthetic data, and 2) show how we can gain insights into the parameter space by focusing on streamline acceptance using real-world data. We demonstrate the potential added value of SSP to the current state of tractography by showing how SSP can be used to reveal patterns in the parameter space.

Keywords: Probabilistic Tractography · Streamline-specific parameters · Parameter Space Exploration

1 Introduction

Diffusion MRI is a non-invasive, in vivo imaging technique that can probe the underlying white matter architecture using tractography. Most tractography algorithms [6,9–11] can be classified as either deterministic or probabilistic [19]. In both cases, many parameters (such as step size, angle, or tracking thresholds) have to be manually set (and fixed across subjects and bundles) beforehand. However, due to anatomical heterogeneity, it is sometimes necessary to adjust these parameters at the individual level, especially in pathological cases [12,14]. Machine learning approaches have been proposed in healthy brains [25] and more recently for diseased brains [23]. The commonality of existing methods puts a focus on defining bundle-specific parameters (BSP) to improve tracking results [5,21]. The main challenge arises when the *generally accepted* parameters start to fail. A potential solution would be to look at streamline-specific parameters (SSP). Here, we explore the viability of SSP in existing tractography algorithms with the main goal of improving tracking results. We use an in-house tracking algorithm based on general probabilistic tractography algorithms that allows the per-streamline parameter sampling and the extraction of the information required for our analysis.

Ensemble tractography [12,13] is a similar method that was proposed to solve the same issues. However, with ensemble tractography N sets of parameters are used to generate N tractograms – each with many streamlines – which are merged at a later stage and an optimised set of parameters is obtained. SSP takes M samples of a given parameter space and generates M streamlines, resulting in a much denser sampling of the parameter space. Using SSP to obtain insights (e.g. optimised parameter settings) proves to be a complex task due to the vast number of parameters and the extensive resulting parameter space. Parameters can be categorized as numerical (e.g. step size, maximum angle between successive steps, number of samples) or spatial (e.g. seed regions, tracking masks, inclusion regions). Aside from these parameters, there are also *rule-based* parameters that are used to decide whether to accept or reject a generated streamline (e.g., filtering parameters). In practice, each algorithm has its own additional set of parameters, and that is without taking into account parameters related to anatomical priors [16,17,20,21] .

In this study, we present an analysis of a subset of the multi-dimensional parameter space to explore the viability of future automatic exploration methods. An in-house implementation of a tracking algorithm is validated and used for the analysis.

2 Methods

2.1 Tracking Algorithm

Streamline specific parameter tracking (SSPT) is a probabilistic tracking method similar to iFOD2 as implemented in MRtrix3 [6,7] and is intended to serve as a representation of probabilistic tracking algorithms, relying only on the Fiber Orientation Distribution functions (FODs) [15] and input ROIs. At a given position, SSPT picks a random direction $\mathbf{d'}$ in a cone around the current tracking direction $\mathbf{d}$. Much like iFOD2, SSPT then calculates an approximation of the joint probability of a step in direction $\mathbf{d'}$ by extrapolating a path in that direction and evaluating the FODs along the path using a discretized sphere and precomputed spherical harmonics. SSPT does this by simply sampling 4 points linearly between $\mathbf{p}$ and $\mathbf{p'}$, whereas iFOD2 places these points along a circular arc such that the current direction is tangent to the start of the arc, and the sampled direction is tangent to the end of the arc. SSPT then chooses a direction $\mathbf{d'}$ at random out of at most 4 potential directions[1] where directions that do not pass the FOD threshold are not considered. Each direction is weighted by the approximation of the joint probability. The choice was made to use this approach over rejection sampling like in iFOD2 because of the biases rejection sampling can add in the interpretation of the parameter distributions. For example, one set of parameters might need many rejections to find a specific trajectory, whereas

---

[1] The exception to this is the very first direction, since this is sampled over the entire sphere. A fixed 32 samples are taken instead to increase the chances a valid starting direction is found when possible.

Algorithm 1. Streamline Specific Parameter Tracking

Input: Random seed, inclusion/exclusion ROIs, FODs
Output: Tracking information for 1 seed and random set of parameters.
1: Generate random set of parameters
2: points ← {seed}
3: Let **d** be a valid start direction if possible, otherwise terminate
4: **while** |points| · step_size < max_length **do**
5:     candidates ← ∅
6:     weights ← ∅
7:     **repeat** n_samples **times**   ▷ From Table 1: n_samples = 4
8:         Sample random direction **d'** in cone around **d**.
9:         Let x_i be points along straight path from **p** to **p'** := **p** + step_size · **d'**
10:        **if** $\forall x_i$, eval_fod(x_i, **d'**) > fod_threshold **then**
11:           Add **d'** to candidates and add $\prod_{x_i} x_i$ to weights.
12:        **end if**
13:     **end**
14:
15:     **if** |candidates| > 0 **then**
16:         Take step in new direction sampled from candidates weighted by weights
17:     **else**
18:         Backtrack or terminate
19:     **end if**
20:
21:     Update status regarding inclusion regions
22:
23:     **if** Current point in exclusion region **then**
24:         Backtrack or terminate
25:     **end if**
26: **end while**
27: **return** tracked points, generated parameters, and other tracking information

another set of parameters could find the same trajectory with far fewer samples. Solving this in post-processing would slow down the exploration too much for practical use. Pseudocode outlining SSPT is given in Algorithm 1. Furthermore, SSPT can make use of binary masks for seeding, an arbitrary number of inclusion zones (both 'and' and 'or'), and exclusion zones. Additionally, backtracking was also added as done in [16,17]. The number of times it tries to backtrack is capped at a constant value (see Table 1), and works by simply taking a single step back (instead of forwards) on the currently tracked path.

Finally and most importantly, SSPT allows for tracking each streamline with its own set of parameters sampled from user specified distributions. The output of a single iteration of the algorithm is a data structure containing information about the tracking such as which parameters were used, whether a streamline was found, how many backtracking attempts were performed, and how long it took to compute. Additionally, flags describing the reason for failing are included as well. It is important to note that a streamline is not necessarily part of this data structure, since SSPT can complete without producing a streamline – and does so more often than not by design. The information obtained in this way is used for the analysis of our experiments. The specific parameter distributions used for each experiment are discussed in their corresponding sections.

Parameters. The SSPT algorithm has a few parameters that are kept fixed for every experiment (see Table 1). Those were pre-determined and deliberately kept the same to reduce the size of the parameter space. Additionally, the target number of streamlines to find was set per experiment. This study focuses on two varying parameters: the step size and the radius of curvature. The radius of curvature is sampled uniformly rather than the angle of the cone, since a larger step size requires a larger angle to produce the same curvature. The relation of radius of curvature r, step size Δx, and angle α is defined as follows:

$$r = \frac{\Delta x}{\sin \frac{\alpha}{2}}. \tag{1}$$

The cone angle is calculated from this relation after uniformly sampling between minimum and maximum radius.

Table 1. Fixed tracking parameters over all experiments

Parameter name	Value	Description
SH_resolution	4	Number of subdivisions of discretized sphere.
backtrack_lim	64	Maximum number of times backtracking occurs.
intermediate_steps	4	Number of steps for computing joint probabilities.
n_samples	4	Number of directions that are always sampled.
seed_samples	32	Number of directions sampled at the seed point.
FOD_threshold	0.1	Minimum FOD value in a given direction for a valid step.

Automatic Validation of Tracking Algorithm. For the validation of the tracking algorithm, preprocessed data from the ISMRM2015 challenge's recent update [1] was used. Bidirectional tracking on the reconstructed FODs was performed with seeding in a white matter mask with the grey matter white matter interface as inclusion region. Two million streamlines were generated using a step size between 0.4 mm and 0.6 mm and a radius of curvature between 0.75 mm and 1.0 mm. These ranges were selected after a single test run with larger ranges, where these parameter ranges showed a good balance between high acceptance rate and speed. An FOD amplitude threshold of 0.1 was used. The updated checker script provided by Renauld et al. [4] was then used to score the resulting tractogram for comparison purposes. Our method used the same masks as the 'WM seeding + PFT tracking' entry and ran on all 24 cores of an Intel i7-13700K with 64 GB of RAM.

2.2 Streamline-Specific Experiments

The data used for the streamline-specific experiments consisted of one healthy subject from the minimally preprocessed Human Connectome Project (HCP) [18], and two clinical datasets provided by the Elisabeth TweeSteden Hospital (ETZ) in Tilburg, The Netherlands. The clinical datasets used were preprocessed using by the tractography pipeline currently in use at ETZ [5]. The FODs and masks generated by the pipeline are directly used with no further processing.

Patient A has a Glioblastoma Multiforme, WHO grade 4, in the vicinity of the angular gyrus near the arcuate fasciculus (AF). **Patient B** has an oligodendroglioma, WHO grade 2, in the postcentral gyrus near the corticospinal tract (CST). Both datasets were acquired with a Philips Achieva 3T MRI scanner ($b = 1500$ s/mm^2, 50 diffusion-weighting directions, six b0 s/mm^2 images, 2 mm isotropic voxel size, TE/TR/echo spacing 87/8000/0.2 ms). The HCP dataset was acquired at a 1.25 mm isotropic voxel size, and only the b0 and $b = 3000$ s/mm^2 were used, in 90 directions.

Per-Streamline Parameter Distributions. The following experiment shows how SSPT can provide detailed information during the tracking process of a specific bundle in cases where the fixed-parameter method presents shortcomings. This means that either 1) not enough streamlines can be found within the allowed number of seeds or time; or that 2) parts of the bundle that are expected to be tracked are not present in the output. Two specific bundles were selected to highlight the usefulness of SSPT: the corticospinal tract (CST), and the arcuate fasciculus (AF). The CST is used to show how, within a bundle, streamline specific parameters can be used to target specific parts of the bundle (e.g., fanning) in a healthy subject. To do so, the bundle is segmented into sub-parts using Quickbundles [8]. The step size is sampled uniformly between 0.2 and 2 times the voxel size (i.e., 1.25 mm^3). Going larger than 2 times the voxel size can cause overshoot. The lower bound is in line with literature [22]. The radius of curvature is sampled uniformly between 2 mm and 100 mm. Lower radii result in streamlines that are no longer anatomically viable, and any radii greater than 100mm all end up with many of the same – almost straight – streamlines.

Clinical Datasets. To assess whether the method can be applied in pathological cases, the AF in Patient A and the CST in patient B are reconstructed using SSPT. The results of SSPT are compared with iFOD2 [6] (manual specification of bundle-specific parameters) and TractSeg using Tract Orientation Map tracking [25] (automated approach). For iFOD2 the bundle-specific parameters as determined by Meesters et al. [5] were used and for TractSeg the default parameters were used. In the end, we qualitatively evaluate how tracking is affected by the tumor region in all three approaches. Light filtering (<1% of streamlines) based on fibre-to-bundle coherence [26] was applied to the results of iFOD2 for visualization purposes only.

3 Results

3.1 Validation of Tracking Algorithm

The SSPT algorithm was validated using the ISMRM2015 challenge. The scores achieved were in line with previous scores and therefore SSPT was used as presented in the rest of the experiments. Table 2 shows the scores of our method compared to the results presented by Renauld et al. [4]. It is important to note that the goal is to show that SSPT can serve as a *representative* for probabilistic tracking methods, and not necessarily as an *alternative*. For this reason the exact same FODs and ROIs were used as Renauld et al. and parameters were not optimised to maximise score.

Table 2. ISMRM2015 scores using the 2023 checker script by Renauld et al. [4]

Order	VB (out of 21)	VS	mean OL	mean ORn	mean F1
WM seeding + local tracking	19	**42.0%**	**82.6%**	121.3%	51.7%
WM seeding + PFT tracking	19	33.5%	68.7%	54.1%	58.2%
Interface seeding + PFT tracking	19	28.8%	64.9%	44.5%	**58.4%**
SSPT (Ours)	**20**	32.2%	55.8%	**33.9%**	56.2%

3.2 Parameter Space Exploration

The results of tracking the left CST in the healthy HCP dataset are shown in Fig. 1. Figure 2 shows the distribution of parameters among all accepted streamlines, and Fig. 3 shows the distribution of parameters for three clusters. By choosing parameters with high success rate (i.e., 3 mm for the radius of curvature and 0.625 mm for the step size), the running time is reduced from 3 min to 20 s. Additionally, Fig. 3 shows a clear difference between a cluster that does not seem to have a clear bias (cluster 1), and two clusters that have opposing biases (clusters 2 and 5).

3.3 Clinical Datasets

In pathological cases, it is important that our choice of parameters strikes a balance between false negatives (e.g., too little streamlines) and false positives (e.g., generating too many spurious streamlines) in the vicinity of the tumor.
Patient A: Figure 4 shows tracking results of the right AF in patient A. As shown, iFOD2 and SSPT produce a larger volume of bundles than TractSeg. However, SSPT seems to have a higher coverage of endpoint fanning than TractSeg, without the added spurious streamlines of iFOD2. More specifically, in the circle area is a deflection of the bundle, which is less pronounced in TractSeg than in the tracking of iFOD2 and SSPT.

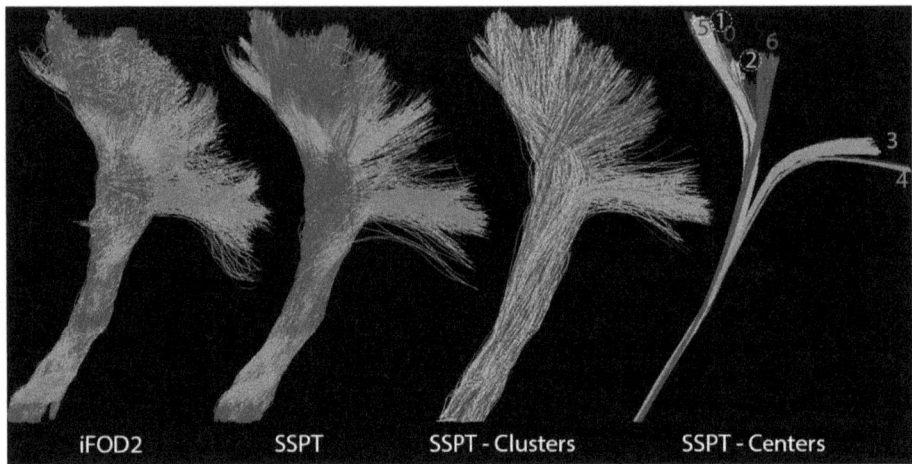

Fig. 1. Trackings of the left CST of the healthy HCP subject. From left to right; results from iFOD2 with the default parameters for the pipeline [5]. Then the result of SSPT with the same ROIs. The last two images show a representation of the cluster representatives. The circled cluster numbers correspond to the data shown in Fig. 3.

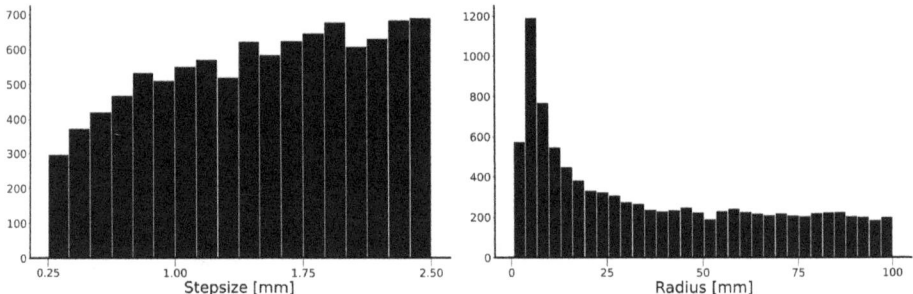

Fig. 2. Histograms of the amount of streamlines out of 10000 that were accepted with a specific parameter in the left CST of the healthy HCP subject.

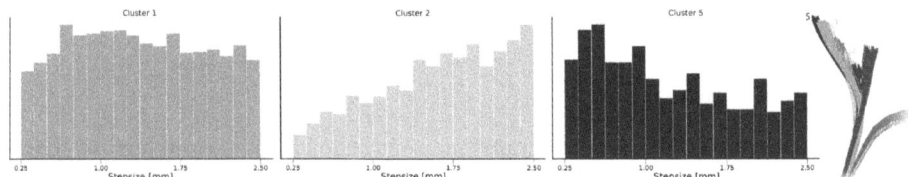

Fig. 3. Histograms for specific clusters as shown in Fig. 1 (left CST of healthy HCP subject). On the right a close up of the cluster representatives is shown.

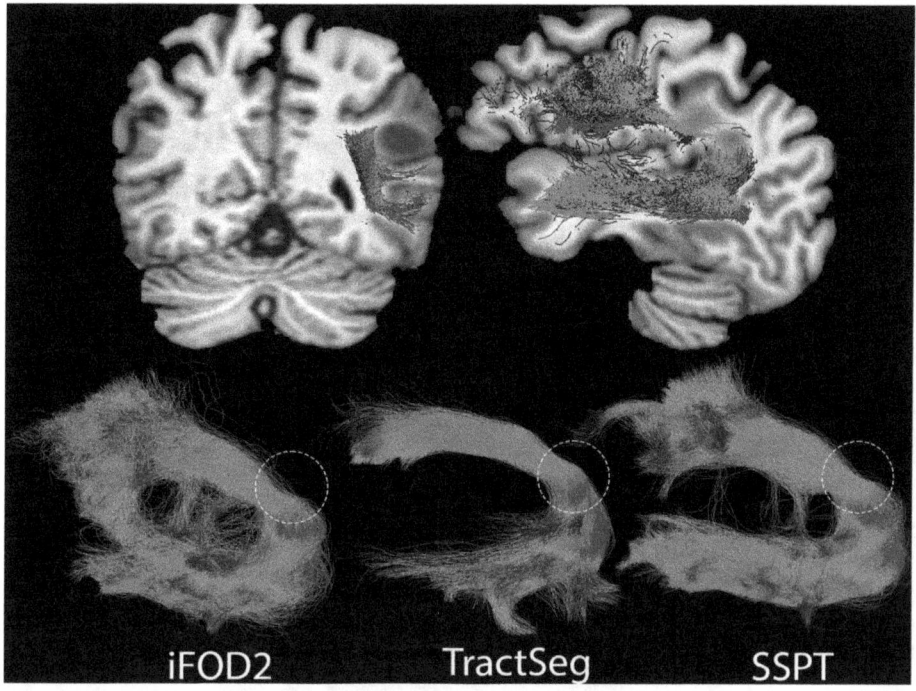

Fig. 4. All three slices are from the T1 image of patient A with a cross-sectional slab of the tracks of the right AF generated by iFOD2 (red), TractSeg (blue), and SSPT (green). The bottom right shows each result again in a 3D view, with a dashed circle highlighting the location of the tumor. (Color figure online)

Patient B: Figure 5 shows the results of tracking the left CST in patient B. One can observe that iFOD2 and SSPT produce a much larger bundle, whereas TractSeg does not produce streamlines that are near the tumor (circled areas). Both iFOD2 and SSPT show curved lines around the tumor, but SSPT produces less spurious streamlines in the tumor area.

4 Discussion

4.1 Experiments

Parameter Space Exploration. The CST of the HCP dataset shows how the parameter distribution can directly influence what subparts of a bundle are more common in terms of occurrence. By looking at the resulting distributions (Fig. 2) of the parameters used to track the entire bundle, one can see that lines with a radius of curvature corresponding to about 3 mm generally have a larger chance of resulting in an accepted streamline. This same stark contrast is not present in the step size distribution. However, since the larger the step size, the greater the chance of acceptance, the user may be inclined to use as a big of a step size as possible, but this could result in anatomically infeasible streamlines. This does show that there is not only computational time to be gained by taking larger steps, but for any given seed the resulting streamline is also more likely to be accepted. The histograms can therefore also be used to selectively adapt the parameter sampling ranges for the respective parameters and increase the likelihood of a seed resulting in an accepted streamline. Doing so speeds up the computation significantly by sampling fewer parameters that have a low probability of finding an accepted streamline. Additionally, Fig. 3 shows that even though clusters 1 and 5 are spatially very close, their resulting histograms are not. Therefore it is possible to target specific parts of the bundle by analysing the histograms per cluster.

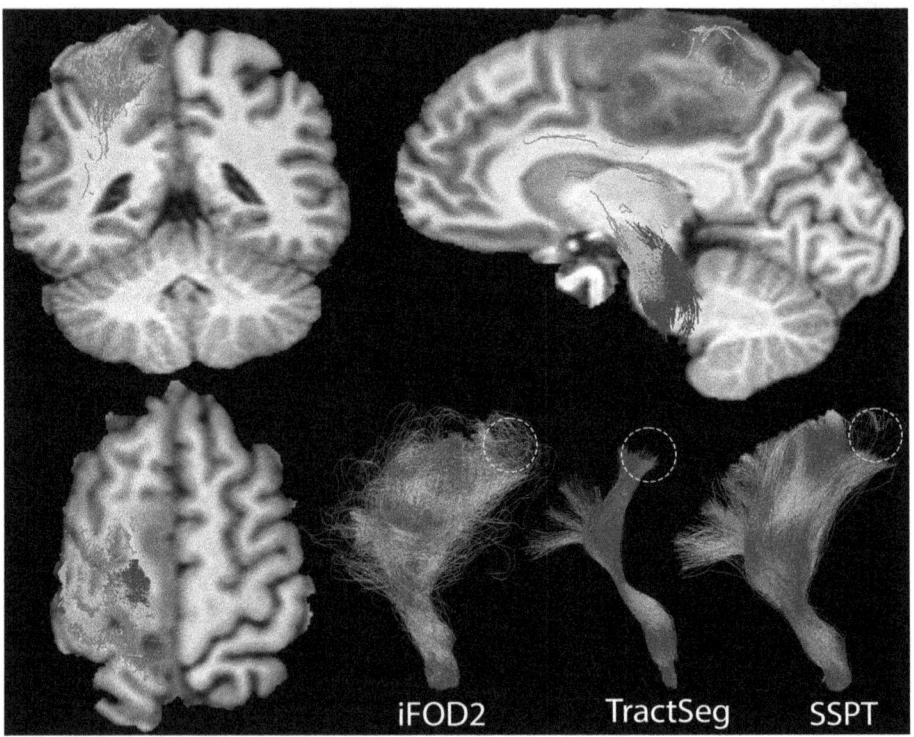

Fig. 5. The top row shows sagittal and coronal slices of the T1 image of patient B with a cross-sectional slab of the tracks of the left CST generated by iFOD2 (red), TractSeg (blue), and SSPT (green). The bottom row shows the same tracks in a 3D view with a dashed circle highlighting the part of the bundle nearest to the tumor. (Color figure online)

Clinical Datasets. Figure 4 showed that iFOD2 seemed to produce more spurious streamlines than both TractSeg and SSPT. This is most likely due to the lower FOD threshold used by iFOD2 for the AF. TractSeg and SSPT both use the same FOD threshold. Therefore it is interesting to investigate the FOD threshold as a variable parameter as well. In Fig. 5, TractSeg misses a part of the bundle near the tumor area that the tracks of iFOD2 and SSPT do contain. This is most likely due to the fact that TractSeg is trained on a dataset of healthy subjects and therefore expects normal anatomy. This is also a plausible explanation for the less pronounced deflection shown in Fig. 4. Using streamline-specific parameters to explore the parameter space would therefore be a viable alternative to deep learning methods trained on healthy patients in pathological cases, since—even with wide parameter ranges being used—SSPT behaves the same as iFOD2 with respect to the two tumors presented.

4.2 Potential Applications

A possible use for parameter space exploration is simplifying the use of tracking algorithms in the clinic since the parameters no longer need to be selected by an expert user, and instead, the domain knowledge of the clinician could suffice.

In the future it would be interesting to step away from spatial clustering and look at similarity between streamlines on a feature level [24]. On top of that, streamline-specific parametrization could facilitate user-guided real-time tracking by presenting the user with different tracking results and adjusting the parameter ranges on the fly based on their input. Additionally, it might be possible to compare the distributions of the parameters between the left and right side of the brain and see whether the successful parameter combinations possibly contain information regarding abnormalities. The choice of tracking algorithm could also be considered a parameter, and ROI-based parameters could be altered using affine transformations or morphological operators. Experimentation with other tracking algorithms and parameters will bring further insight into the effectiveness of SSP.

5 Conclusion

We have shown that parameter space exploration is a viable method of making more informed parameter selections for tractography. Additionally, SSPT can find specific parameter combinations that work well for specific parts of certain bundles if an adequate clustering is provided. This means parameters can be selected with the intent of finding more (or less) of a specific portion of the bundle. A concomitant advantage is that SSPT can be used to speed up tracking by selecting specific parameters. The downside of this approach is that existing algorithms have to be adapted to take SSP as input, and output the information required for analysis. Additionally, SSPT, as presented, does not consider anatomical plausability when sampling the provided parameter distributions. Even though we have shown two cases in which it behaves the same as iFOD2, this is no guarantee that it will always be correct.

Acknowledgements. This publication is part of the project Bringing Tractography into Daily Neurosurgical Practice with project number KICH1.ST03.21.004 of the research programme Key Enabling Technologies for Minimally Invasive Interventions in Healthcare, which is (partly) financed by the Dutch Research Council (NWO). We would like to thank neurosurgeon Geert-Jan Rutten and scientific programmer Rembrandt Bakker for sharing and preprocessing the clinical dataset used in our experiments at the Elisabeth TweeSteden Hospital (ETZ) in Tilburg, The Netherlands. The authors thank Tom Hendriks for methodological support.

References

1. Maier-Hein, K.H., et al.: The challenge of mapping the human connectome based on diffusion tractography. Nat. Commun. **8**(1), 1–13 (2017). https://doi.org/10.1038/s41467-017-01285-x
2. Côté, M.-A., Boré, A., Girard, G., Houde, J.-C., Descoteaux, M.: Tractometer: Online Evaluation System for Tractography, MICCAI (2012). https://doi.org/10.1007/978-3-642-33415-3_86
3. Côté, M.-A., Girard, G., Boré, A., Garyfallidis, E., Houde, J.-C., Descoteaux, M.: Tractometer: towards validation of tractography pipelines. Med. Image Anal. **17**(7), 844–857 (2013). https://doi.org/10.1016/j.media.2013.03.009
4. Renauld, E., Théberge, A., Houde, J.-C., Descoteaux, M.: Validate your white matter tractography algorithms with a reappraised ISMRM 2015 tractography challenge scoring system. Sci. Rep. **13**, 2347 (2023). https://doi.org/10.1038/s41598-023-28560-w
5. Meesters, S., Landers, M., Rutten, G.J., et al.: Subject-specific automatic reconstruction of white matter tracts. J. Digit. Imaging **36**, 2648–2661 (2023). https://doi.org/10.1007/s10278-023-00883-0
6. Tournier, J.-D., Calamante, F., Connelly, A.: Improved probabilistic streamlines tractography by 2nd order integration over fibre orientation distributions. In: Proceedings of the International Society for Magnetic Resonance in Medicine (2010)
7. Tournier, J.-D., et al.: MRtrix3: a fast, flexible and open software framework for medical image processing and visualisation. NeuroImage, 202 (2019). https://doi.org/10.1016/j.neuroimage.2019.116137
8. Garyfallidis, E., Brett, M., Correia, M.M., Williams, G.B., Nimmo-Smith, I.: QuickBundles, a method for tractography simplification. Front. Neurosci. **11**(6), 175 (2012). https://doi.org/10.3389/fnins.2012.00175
9. Mori, S., Crain, B.J., Chacko, V.P., van Zijl, P.C.M.: Three-dimensional tracking of axonal projections in the brain by magnetic resonance imaging. Ann. Neurol. **45**, 265–269 (1999). https://doi.org/10.1002/1531-8249(199902)45:2<265::aid-ana21>3.0.co;2-3
10. Jones, D.: Tractography gone wild: probabilistic fibre tracking using the wild bootstrap with diffusion tensor MRI. IEEE Trans. Med. Imaging **27**, 1268–1274 (2008). https://doi.org/10.1109/TMI.2008.922191
11. Basser, P.J., Pajevic, S., Pierpaoli, C., Duda, J., Aldroubi, A.: In vivo fiber tractography using DT-MRI data. Magn. Reson. Med. **44**, 625–632 (2000). https://doi.org/10.1002/1522-2594(200010)44:4<625::aid-mrm17>3.0.co;2-o
12. Takemura, H., Caiafa, C.F., Wandell, B.A., Pestilli, F.: Ensemble Tractography. PLoS Comput. Biol. (2016). https://doi.org/10.1371/journal.pcbi.1004692

13. Joanisse, A., et al.: Improving white matter bundle recovery: a fast & practical ensemble tractography pipeline. In: 29th Annual Conference & Exhibition ISMRM, Vancouver, Canada, vol. 5 (2021)
14. Chamberland, M., Whittingstall, K., Mathieu, D., Fortin, D., Descoteaux, M.: Real-time multi-peak tractography for instantaneous connectivity display. Front. Neuroinform. **8**, 59 (2014). https://doi.org/10.3389/fninf.2014.00059
15. Tournier, J.D., Calamante, F., Connelly, A.: Robust determination of the fibre orientation distribution in diffusion MRI: non-negativity constrained super-resolved spherical deconvolution. Neuroimage **35**, 1459–1472 (2007). https://doi.org/10.1016/j.neuroimage.2007.02.016
16. Smith, R.E., Tournier, J.D., Calamante, F., Connelly, A.: Anatomically-constrained tractography: improved diffusion MRI streamlines tractography through effective use of anatomical information. Neuroimage **62**(3), 1924–1938 (2012). https://doi.org/10.1016/j.neuroimage.2012.06.005
17. Girard, G., Whittingstall, K., Deriche, R., Descoteaux, M.: Towards quantitative connectivity analysis: reducing tractography biases. Neuroimage (2014). https://doi.org/10.1016/j.neuroimage.2014.04.074
18. Glasser, M.F., et al.: The minimal preprocessing pipelines for the Human Connectome Project." Neuroimage **80**, 105–124 (2013). https://doi.org/10.1016/j.neuroimage.2013.04.127
19. Descoteaux, M., Deriche, R., Knosche, T.R., Anwander, A.: Deterministic and probabilistic tractography based on complex fibre orientation distributions. IEEE Trans. Med. Imaging **28**(2), 269–286 (2008). https://doi.org/10.1109/TMI.2008.2004424
20. Chamberland, M., et al.: Active delineation of Meyer's loop using oriented priors through MAGNEtic tractography (MAGNET). Hum. Brain Mapp. **38**(1), 509–527 (2017). https://doi.org/10.1002/hbm.23399
21. Rheault, F., et al.: Bundle-specific tractography with incorporated anatomical and orientational priors. Neuroimage **186**, 382–398 (2019). https://doi.org/10.1016/j.neuroimage.2018.11.018
22. Tournier, J.D., Calamante, F., Connelly, A.: MRtrix: diffusion tractography in crossing fiber regions. Int. J. Imaging Syst. Technol. **22**(1), 53–66 (2012). https://doi.org/10.1002/ima.22005
23. Peretzke, R., et al.: atTRACTive: semi-automatic white matter tract segmentation using active learning. In: Greenspan, H., et al. Medical Image Computing and Computer Assisted Intervention - MICCAI 2023. LNCS, vol. 14227. Springer, Cham (2023). https://doi.org/10.48550/arXiv.2305.18905
24. Li, Y., Wang, C., Shene, C.-K.: Streamline similarity analysis using bag-of-features. Proc. SPIE Int. Soc. Optical Eng.. **9017**, 90170N (2013). https://doi.org/10.1117/12.2038253
25. Wasserthal, J., Neher, P.F., Hirjak, D., Maier-Hein, K.H.: Combined tract segmentation and orientation mapping for bundle-specific tractography. Med. Image Anal. **58**, 101559 (2019). https://doi.org/10.1016/j.media.2019.101559
26. Meesters, S., Ossenblok, P., Wagner, L., et al.: Stability metrics for optic radiation tractography: towards damage prediction after respective surgery. J. Neurosci. Methods (2017). https://doi.org/10.1016/j.jneumeth.2017.05.029

Cross-Domain Fiber Cluster Shape Analysis for Language Performance Cognitive Score Prediction

Yui Lo[1,2,4], Yuqian Chen[1,2], Dongnan Liu[4], Wan Liu[5], Leo Zekelman[2,7], Fan Zhang[6], Yogesh Rathi[1,2], Nikos Makris[1,3], Alexandra J. Golby[1,2], Weidong Cai[4], and Lauren J. O'Donnell[1,2(✉)]

[1] Harvard Medical School, Boston, USA
[2] Brigham and Women's Hospital, Boston, USA
odonnell@bwh.harvard.edu
[3] Massachusetts General Hospital, Boston, USA
[4] The University of Sydney, Sydney, Australia
[5] Beijing Institute of Technology, Beijing, China
[6] University of Electronic Science and Technology of China, Chengdu, China
[7] Harvard University, Boston, USA

Abstract. Shape plays an important role in computer graphics, offering informative features to convey an object's morphology and functionality. Shape analysis in brain imaging can help interpret structural and functionality correlations of the human brain. In this work, we investigate the shape of the brain's 3D white matter connections and its potential predictive relationship to human cognitive function. We reconstruct fiber clusters as sequences of 3D points using diffusion magnetic resonance imaging (dMRI) tractography. To describe each connection, we extract 12 shape descriptors in addition to traditional dMRI connectivity and tissue microstructure features. We introduce a novel framework, *Shape-Fused Fiber cluster Transformer* (SFFormer), that leverages a multi-head cross-attention feature fusion module to predict subject-specific language performance based on dMRI tractography. We assess the performance of the method on a large dataset including 1065 healthy young adults. The results demonstrate that both the transformer-based SFFormer model and its inter/intra feature fusion with shape, microstructure, and connectivity are informative, and together, they improve the prediction of subject-specific language performance scores compared to conventional models. Overall, our results indicate that the shape of the brain's connections is predictive of human language function.

Keywords: Shape analysis · tractography · diffusion MRI · deep embeddings · domain-fusion

1 Introduction

The study of 3D shape has long been recognized as crucial for computer graphics and medical image analysis [35]. In the field of magnetic resonance imaging

(MRI), the study of shape has enabled detailed analyses of the folding of the brain's cortex and the morphology of subcortical gray matter structures [8]. However, the shape of the brain's white matter connections, which transmit information throughout the brain, has been much less studied.

Diffusion MRI (dMRI) tractography is a unique method that enables the 3D reconstruction of the brain's white matter connections based on water diffusion in brain tissue [2]. dMRI tractography produces sequences of 3D points, called streamlines, which can be grouped to define individual brain connections or fiber clusters that have different anatomical shapes (Fig. 1). Quantitative analyses of fiber clusters include tissue microstructure (using water diffusion in tissue), brain connectivity (strength of each connection), and shape analyses. Measures of shape capture white matter variability across individuals [33] and changes in aging [23]. However, the functional importance of the shape of white matter connections is not well understood. To assess whether fiber cluster shape may be related to language performance, in this work we employ a testbed task of predicting individual language performance. We assess whether the integration of information across shape, microstructure, and connectivity feature domains can enhance the prediction of individual language performance.

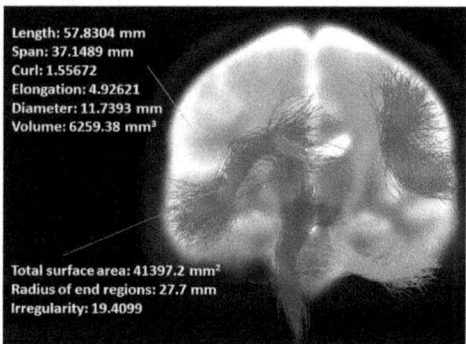

Fig. 1. Four example individual white matter connections (fiber clusters) extracted from the entire white matter of the human brain using a fiber clustering approach [37]. Example shape descriptors are extracted for the blue fiber cluster. (Color figure online)

1.1 Related Work

In this section, we first give an overview of methods that have been proposed for the prediction of individual language function using dMRI tractography data, then we briefly describe the deep learning techniques upon which our current framework is built.

In the literature, several approaches have been proposed to predict individual cognitive and/or language functional performance using dMRI tractography data [5,10,14,18]. Tissue microstructure measures derived from dMRI have been shown to relate to language function using traditional (non-deep learning) regression analysis [34]. The studied measures included the fractional anisotropy (FA), which describes the anisotropy of water diffusion within brain tissue, and the mean diffusivity (MD), which describes the overall magnitude of water diffusion [20], as well as the number of streamlines (NoS), which is thought to relate to the connectivity of the brain [36]. In contrast to these traditional features, fiber clusters can be described by shape measures such as surface area and volume, as well as recently proposed, fiber-tract-specific measures such as the surface area of the region where the tract inserts in to the gray matter [33]. Other shape measures have been proposed for tractography, including curvature and torsion [3], fiber dispersion [22], and volume [17].

Recent deep-learning methods have investigated the prediction of individual language performance using dMRI tractography. A convolutional neural network (CNN) based deep learning method has shown that connectivity is predictive of language proficiency in children with epilepsy [14]. A geometric deep-learning approach showed that microstructure and connectivity are predictive of language function in healthy young adults [5]. A multilayer network approach was applied to predict language performance in human aging [7]. In contrast with these methods, we focus on a novel transformer-based network design.

In recent years, transformer models [11] are increasingly popular for computer vision tasks such as object detection [40], classification [6], and segmentation [25]. The advantage of transformers over CNNs is the use of multi-head self-attention [30] to enhance the model's ability to interpret complex semantic and structural feature relationships more comprehensively. Transformers have also been shown to be successful in many medical image applications, including dMRI [4,27,32,38]. There is a substantial body of literature on transformer models to predict tissue microstructure, including SwinDTI [27], Microstructure Estimation Transformer with Sparse Coding [39], Hybrid Graph Transformer (HGT) [4], and 3D HGT [32]. Applications of transformers in tractography analysis are relatively limited, such as TractoFormer [38] for whole-brain tractography analysis. Consequently, it is of interest to investigate the application of transformers in the analysis of tractography data.

2 Methodology

2.1 Tractography and Fiber Clustering

In this work, we study the shape of the fiber clusters of 1065 healthy young adults (575 females and 490 males, 28.7 years old on average) from the Human Connectome Project Young Adult (HCP-YA) dataset [28,29]. Whole brain tractography is generated for each subject's dMRI data using a two-tensor unscented Kalman filter method [19] that can represent multiple crossing fibers, enabling anatomically sensitive estimation of the pathway and connectivity of fiber clusters [12]. Tractography is then parcellated into 953 fiber clusters using an anatomically curated tractography brain atlas [37]. Each fiber cluster contains hundreds of streamlines and represents a particular connection in the human brain (Fig. 1).

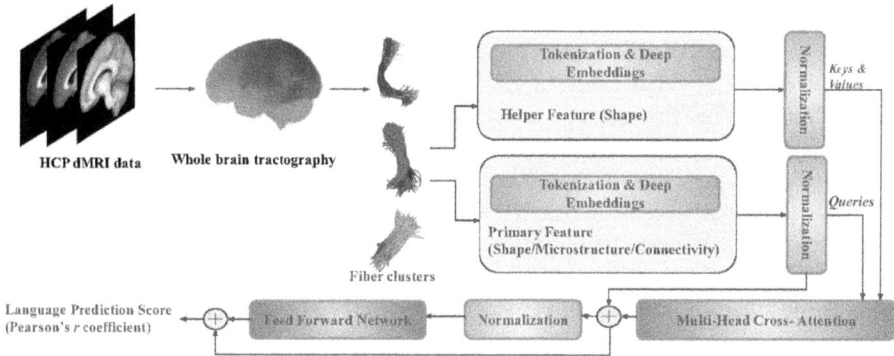

Fig. 2. Overview of the SFFormer framework. HCP dMRI data undergoes whole brain tractography to obtain 953 fiber clusters. The microstructure, connectivity, and shape features of the fiber clusters are calculated and used as inputs to the SFFormer framework that leverages both a helper feature and a primary feature in the multi-head cross-attention module to output a language prediction score.

2.2 Traditional and Shape Features

For each cluster, we compute traditional tissue microstructure features including fractional anisotropy (FA) and mean diffusivity (MD), and the traditional connectivity feature of the number of streamlines (NoS) [36]. These features are used to compare and evaluate the shape features.

We study 12 fiber cluster shape features that are considered to provide a comprehensive shape analysis of tractography [33]. Features include length, diameter, elongation, span, curl, volume, trunk volume, branch volume, total surface area, total radius of end regions, total area of end regions, and irregularity. These shape features are computed for all fiber clusters from all subjects by applying the software DSIStudio [33]. Full definitions and calculations of the shape measures are presented in DSIStudio [33].

2.3 Shape-Fused Fiber Cluster Transformer (SFFormer)

In this section, we present our proposed SFFormer for subject-specific language score prediction. As depicted in Fig. 2, the SFFormer model comprises a tokenization module and an encoder-only transformer architecture, specifically tailored for prediction tasks. This encoder-only design aligns with the task of focusing on learning the fiber clusters' features for language score predictive outcomes. The SFFormer encoder comprises a stack of 1-4 identical layers. Each layer includes a multi-head attention module and a feed-forward network.

The tokenization module [11] performs deep embedding of a particular feature (e.g., FA or length) of dimension 1×953. To create the embedding, we multiply the input data (x) with random initialized weights and then add random initialized biases. This process prepares the data for the multi-head cross-attention module in the deep learning pipeline.

We extended our design from the vanilla transformer [11]. We naturally take a fiber cluster feature as a token to utilize the long-range dependency of all cluster features to benefit prediction. We employ a multi-head mechanism [30] that is well suited for processing long sequences, such as the 953 fiber cluster features, because each head independently attends to different parts of the input sequence.

We design a multi-head cross-attention module to fuse features from shape, microstructure, and connectivity feature domains. Instead of using the transformer's self-attention mechanism, our multi-head cross-attention module can fuse the features of different domains to symmetrically combine two embedding sequences of the same dimension, where one sequence is used as the query (Q) input. The other sequence is used as the key (K) and value (V) inputs in SFFormer to provide feature fusion. As it requires two embeddings SFFormer captures and attends to information from different features simultaneously. The motivation is to more effectively determine varying attention weights by utilizing the dual-stream input framework. This methodology emphasizes cross-attention to concurrently train on the primary feature to attempt to integrate key information from both data streams.

2.4 Implementation Details

Our model is trained and tuned with Optuna Hyperparameters [1], set to 20 trials. The model is configured with the ReGLU activation and the He initialization [13] with 8 attention heads. The model is trained and evaluated with batch sizes of 8 for 1000 epochs with patience of 50 epochs. All of the experiments are split into three-fold cross-validation. The training is optimized with Adam [16], where the learning rate is set between 1e-5 and 1e-3 with a log uniform weight decay between 1e-6 and 1e-3. The tokens are set between 64 to 512 with larger embeddings capturing more information. The dropouts for attention and feed-forward modules are set between 0 and 0.5, and 0 and 0.2 for residual connections. All experiments are conducted on an NVIDIA RTX A5000 GPU using PyTorch 1.7.1 [21].

3 Experiments and Results

First, we conduct experiments to perform subject-specific language score prediction based on individual features. We compare the performance of a state-of-the-art 1DCNNN model [18] and a baseline transformer model, when trained on an individual microstructure, connectivity, or shape feature. Next, we fuse each feature with a selected helper shape feature and apply the SFFormer model.

The helper shape feature is selected as the best-performing shape feature when using the baseline transformer model. We select diameter as the helper shape feature for TPVT score prediction and irregularity as the helper shape feature for TORRT score prediction.

3.1 Language Assessments Scores

We predict subject-specific performance on two language assessments provided by HCP-YA, including the NIH Toolbox Picture Vocabulary Test (TPVT) and the NIH Toolbox Oral Reading Recognition Test (TORRT) [9,31]. TPVT measures vocabulary comprehension and is a receptive language assessment [9]. TORRT measures reading decoding and is a spoken language assessment [9].

3.2 Evaluation Metric

The Pearson correlation coefficient (r) [24] is employed to evaluate language performance prediction as it is a prevalent metric in neurocognitive performance prediction [10,15,26]. Pearson's r measures the strength and direction (positive or negative) of the linear association between two variables.

3.3 Results and Discussions

Tables 1 and 2 show the performance of the three compared models for predicting subject-specific vocabulary comprehension (TPVT) and subject-specific oral reading (TORRT) scores, respectively.

The CNN model [18], shown in the second column of Tables 1 and 2, successfully predicts language performance, though it is outperformed by both the baseline transformer and SFFormer models. When using the CNN model, the NoS feature is the most informative traditional feature, while several shape features (shown in italics) outperform NoS.

The baseline transformer model (third column of Tables 1 and 2) outperforms the CNN model for all input features. This indicates that the transformer improves the performance of the language score prediction task. The FA feature is the most informative traditional feature. Multiple shape features (shown in italics) outperform FA, including diameter (Table 1), volume, diameter, total surface area, and irregularity (Table 2).

The SFFormer (fourth column of Tables 1 and 2) successfully predicts language performance, and most of its features outperform the baseline model as well as the state-of-the-art CNN model. This indicates that the domain fusion technique effectively contributes to subject-specific language score prediction, where the performance improves when information from the helper feature is included to aid the overall training. In Table 1, shape features have comparable

Table 1. Prediction performance for TPVT (r). Shape features shown in italics outperform the best-performing traditional feature. Bolded features show the best performance across the three different learning models.

Features	CNN [18]	Vanilla Transformer	SFFormer (helper: diameter)
Microstructure			
FA	0.293±0.063	**0.418±0.077**	0.404±0.079
MD	0.260±0.041	0.337±0.098	**0.338±0.098**
Connectivity			
NoS	0.395±0.054	0.410±0.103	**0.417±0.007**
Shape			
Length	0.133±0.039	0.330±0.079	**0.414±0.080**
Span	0.119±0.044	0.355±0.094	**0.417±0.098**
Curl	0.203±0.092	0.310±0.070	**0.407±0.081**
Volume	0.381±0.063	0.410±0.102	*0.423±0.071*
Trunk Volume	0.156±0.083	0.275±0.041	**0.414±0.084**
Branch Volume	0.376±0.064	0.414±0.096	*0.430±0.079*
Diameter	0.406±0.082	**0.419±0.083**	—
Elongation	0.313±0.070	0.392±0.074	*0.419±0.083*
Total surface area	0.395±0.060	**0.418±0.098**	0.406±0.092
Radius of end regions	0.235±0.045	0.347±0.125	*0.429±0.084*
Surface area of end regions	0.406±0.080	0.414±0.100	*0.418±0.092*
Irregularity	0.322±0.041	0.391±0.092	**0.416±0.071**

or better performance than FA and MD, with various shape features (shown in italics), including the surface area of end regions, elongation, volume, radius of end regions, and branch volume, surpassing the traditionally best-performing feature, NoS, in predicting language performance. Also, Table 2 reveals that FA is the most informative traditional feature, and FA outperforms several shape features, such as trunk volume and diameter (shown in italics).

In summary, the evaluation presented in Tables 1 and 2 demonstrates the superior predictive power of shape features and domain fusion in the SFFormer model, marking an improvement over traditional features and surpassing the state-of-the-art methods of the CNN model and the self-attention vanilla transformer model. Future work may explore models that incorporate multiple input features.

Table 2. Prediction performance for TORRT (r). Shape features shown in italics outperform the best-performing traditional feature. Bolded features show the best performance across the three different learning models.

Features	CNN [18]	Vanilla Transformer	SFFormer (helper: irregularity)
Microstructure			
FA	0.332±0.055	0.382±0.059	**0.383±0.06**
MD	0.315±0.004	0.344±0.021	**0.374±0.06**
Connectivity			
NoS	0.349±0.024	0.345±0.061	**0.372±0.05**
Shape			
Length	0.103±0.002	0.301±0.056	**0.376±0.053**
Span	0.126±0.017	0.318±0.071	**0.377±0.072**
Curl	0.241±0.014	0.285±0.061	**0.377±0.075**
Volume	0.324±0.016	*0.392±0.083*	0.379±0.066
Trunk Volume	0.184±0.035	0.260±0.039	*0.384±0.123*
Branch Volume	0.357±0.021	**0.377±0.075**	0.362±0.073
Diameter	0.315±0.038	0.390±0.071	*0.398±0.050*
Elongation	0.275±0.005	0.363±0.045	*0.376±0.049*
Total surface area	0.368±0.046	**0.391±0.079**	0.369±0.056
Radius of end regions	0.3196±0.063	0.341±0.087	**0.374±0.053**
Surface area of end regions	0.330±0.001	**0.374±0.085**	0.371±0.062
Irregularity	0.341±0.021	*0.439±0.062*	—

4 Conclusion

In this paper, we proposed the SFFormer, which utilizes a multi-head cross-attention module to fuse features from different domains to improve the prediction results. Our SFFormer results show that measures of the shape of fiber cluster connections are informative for the prediction of individual, subject-specific language performance. The evaluation of the HCP-YA dataset suggests inter/intra domain feature fusion to be beneficial towards better prediction. This suggests that shape-related features are useful for predicting and evaluating various cognitive abilities, potentially outperforming microstructural and connectivity features in certain scenarios. Overall, this suggests that the shape of the white matter fiber clusters relates to important functions of the human brain.

Acknowledgements. This work is in part supported by the National Key R&D Program of China (No. 2023YFE0118600), and the National Natural Science Foundation of China (No. 62371107). This work is supported by the University of Sydney International Scholarship

Compliance with Ethical Standards. This study uses public HCP imaging data; no ethical approval was required.

References

1. Akiba, T., Sano, S., Yanase, T., Ohta, T., Koyama, M.: Optuna: a next-generation hyperparameter optimization framework. In: Proceedings of the 25th ACM SIGKDD International Conference on Knowledge Discovery & Data Mining, pp. 2623–2631. KDD '19, Association for Computing Machinery, New York, NY, USA (2019)
2. Basser, P., Pajevic, S., Pierpaoli, C., Duda, J., Aldroubi, A.: In vivo fiber tractography using DT-MRI data. Magn. Reson. Med. **44**(4), 625–632 (2000)
3. Batchelor, P.G., Calamante, F., Tournier, J.D., Atkinson, D., Hill, D., Connelly, A.: Quantification of the shape of fiber tracts. Magn. Reson. Med. **55**(4), 894–903 (2006)
4. Chen, G., et al.: Hybrid graph transformer for tissue microstructure estimation with under sampled diffusion MRI data. In: Medical Image Computing and Computer Assisted Intervention – MICCAI 2022, pp. 113–122. Springer Nature Switzerland (2022)
5. Chen, Y., et al.: TractGeoNet: a geometric deep learning framework for pointwise analysis of tract microstructure to predict language assessment performance. arXiv preprint arXiv:2307.03982 (2023)
6. Dosovitskiy, A., et al.: An image is worth 16×16 words: transformers for image recognition at scale. arXiv preprint arXiv:2010.11929 (2020)
7. Feng, G., et al.: Methodological evaluation of individual cognitive prediction based on the brain white matter structural connectome. Hum. Brain Mapp. **43**(12), 3775–3791 (2022)
8. Fischl, B.: FreeSurfer. Neuroimage **62**(2), 774–781 (2012)
9. Gershon, R., et al.: Language measures of the NIH toolbox cognition battery. J. Int. Neuropsychol. Soc. **20**(6), 642–651 (2014)
10. Gong, W., Beckmann, C., Smith, S.: Phenotype discovery from population brain imaging. Med. Image Anal. **71**, 102050 (2021)
11. Gorishniy, Y., Rubachev, I., Khrulkov, V., Babenko, A.: Revisiting deep learning models for tabular data. Adv. Neural. Inf. Process. Syst. **34**, 18932–18943 (2021)
12. He, J., et al.: Reconstructing the somatotopic organization of the corticospinal tract remains a challenge for modern tractography methods. Hum. Brain Mapp. **44**(17), 6055–6073 (2023)
13. He, K., Zhang, X., Ren, S., Sun, J.: Delving deep into rectifiers: Surpassing Human-Level performance on ImageNet classification. In: 2015 IEEE International Conference on Computer Vision (ICCV), pp. 1026–1034. IEEE (2015)
14. Jeong, J., Lee, M., O'Hara, N., Juhász, C., Asano, E.: Prediction of baseline expressive and receptive language function in children with focal epilepsy using diffusion tractography-based deep learning network. Epilepsy Behav. **117**, 107909 (2021)
15. Kim, M., et al.: A structural enriched functional network: an application to predict brain cognitive performance. Med. Image Anal. **71**, 102026 (2021)
16. Kingma, D., Ba, J.: Adam: a method for stochastic optimization. arXiv preprint arXiv:1412.6980 (2014)
17. Lebel, C., Gee, M., Camicioli, R., Wieler, M., Martin, W., Beaulieu, C.: Diffusion tensor imaging of white matter tract evolution over the lifespan. Neuroimage **60**(1), 340–352 (2012)

18. Liu, W., et al.: Fiber tract shape measures inform prediction of Non-Imaging phenotypes. arXiv preprint arXiv:2303.09124 (2023)
19. Malcolm, J., Shenton, M., Rathi, Y.: Filtered multitensor tractography. IEEE Trans. Med. Imaging **29**(9), 1664–1675 (2010)
20. O'Donnell, L., Westin, C.: An introduction to diffusion tensor image analysis. Neurosurg. Clin. N. Am. **22**(2), 185–96, viii (2011)
21. Paszke, A., et al.: PyTorch: an imperative style, high-performance deep learning library. In: Proceedings of the 33rd International Conference on Neural Information Processing Systems, pp. 8026–8037. Curran Associates Inc., Red Hook, NY, USA (2019)
22. Savadjiev, P., Kindlmann, G.L., Bouix, S., Shenton, M.E., Westin, C.F.: Local white matter geometry from diffusion tensor gradients. Neuroimage **49**(4), 3175–3186 (2010)
23. Schilling, K.G., et al.: Aging and white matter microstructure and macrostructure: a longitudinal multi-site diffusion MRI study of 1218 participants. Brain Struct. Funct. **227**(6), 2111–2125 (2022)
24. Sedgwick, P.: Pearson's correlation coefficient. British Medical J. **345** (2012)
25. Strudel, R., Garcia, R., Laptev, I., Schmid, C.: Segmenter: transformer for semantic segmentation. In: 2021 IEEE/CVF International Conference on Computer Vision (ICCV), pp. 7262–7272. IEEE (2021)
26. Tian, Y., Zalesky, A.: Machine learning prediction of cognition from functional connectivity: are feature weights reliable? Neuroimage **245**, 118648 (2021)
27. Tiwari, A., Singh, R., Shigwan, S.: SwinDTI: swin transformer-based generalized fast estimation of diffusion tensor parameters from sparse data. Neural Comput, Appl. (2023)
28. Van Essen, D., Smith, S., Barch, D., Behrens, T., Yacoub, E., Ugurbil, K.: WU-Minn HCP Consortium: the WU-Minn human connectome project: an overview. Neuroimage **80**, 62–79 (2013)
29. Van Essen, D., et al.: WU-Minn HCP Consortium: the human connectome project: a data acquisition perspective. Neuroimage **62**(4), 2222–2231 (2012)
30. Vaswani, A., et al.: Attention is all you need. In: Advances in Neural Information Processing Systems, vol. 30 (2017)
31. Weintraub, S., et al.: Cognition assessment using the NIH toolbox. Neurology **80**(11 Suppl 3), S54-64 (2013)
32. Yang, J., et al.: Towards accurate microstructure estimation via 3D hybrid graph transformer. In: Medical Image Computing and Computer Assisted Intervention – MICCAI 2023, pp. 25–34. Springer Nature Switzerland (2023)
33. Yeh, F.: Shape analysis of the human association pathways. Neuroimage **223**, 117329 (2020)
34. Zekelman, L., et al.: White matter association tracts underlying language and theory of mind: an investigation of 809 brains from the human connectome project. Neuroimage **246**, 118739 (2022)
35. Zhang, D., Lu, G.: Review of shape representation and description techniques. Pattern Recognit. **37**(1), 1–19 (2004)
36. Zhang, F., et al.: Quantitative mapping of the brain's structural connectivity using diffusion MRI tractography: a review. Neuroimage **249**, 118870 (2022)
37. Zhang, F., et al.: An anatomically curated fiber clustering white matter atlas for consistent white matter tract parcellation across the lifespan. Neuroimage **179**, 429–447 (2018)

38. Zhang, F., Xue, T., Cai, W., Rathi, Y., Westin, C.F., O'Donnell, L.J.: Tracto-Former: a novel fiber-level whole brain tractography analysis framework using spectral embedding and vision transformers. In: Wang, L., Dou, Q., Fletcher, P.T., Speidel, S., Li, S. (eds.) Medical Image Computing and Computer Assisted Intervention - MICCAI 2022, pp. 196–206. Springer Nature Switzerland, Cham (2022)
39. Zheng, T., et al.: A microstructure estimation transformer inspired by sparse representation for diffusion MRI. Med. Image Anal. **86**, 102788 (2023)
40. Zhu, X., Su, W., Lu, L., Li, B., Wang, X., Dai, J.: Deformable DETR: deformable transformers for End-to-End object detection. arXiv preprint arXiv:2010.04159 (2020)

Can Transfer Learning Improve Supervised Segmentation of White Matter Bundles in Glioma Patients?

Chiara Riccardi[1,2](✉), Sofia Ghezzi[2], Gabriele Amorosino[1,2], Luca Zigiotto[3], Silvio Sarubbo[3], Jorge Jovicich[2], and Paolo Avesani[1,2]

[1] Neuroinformatics Laboratory, Fondazione Bruno Kessler, Trento, Italy
[2] Center for Mind/Brain Sciences, University of Trento, Trento, Italy
chiara.riccardi-1@unitn.it
[3] Neurosurgery Department, Azienda Provinciale Servizi Sanitari, Trento, Italy

Abstract. In clinical neuroscience, the segmentation of the main white matter bundles is propaedeutic for many tasks such as pre-operative neurosurgical planning and monitoring of neuro-related diseases. Automating bundle segmentation with data-driven approaches and deep learning models has shown promising accuracy in the context of healthy individuals. The lack of large clinical datasets is preventing the translation of these results to patients. Inference on patients' data with models trained on the healthy population is not effective because of domain shift. This study aims to carry out an empirical analysis to investigate how transfer learning might be beneficial in overcoming these limitations. For our analysis, we consider a public dataset with hundreds of individuals and a clinical dataset with glioma patients. We focus our preliminary investigation on the corticospinal tract and on the inferior longitudinal fasciculus. The results show that transfer learning is effective in overcoming part of the domain shift.

Keywords: White Matter Bundle Segmentation · Glioma · Deep learning · Transfer learning

1 Introduction

This work aims to carry out an empirical study to investigate the effectiveness of the translation of a supervised learning model for white matter bundle segmentation from a healthy population to patients with glioma.

The segmentation of bundles in patients with glioma provides neurosurgeons a reference for the eloquent regions to be preserved to reduce subsequent cognitive impairment [19].

Manual bundle segmentation is time-consuming. For this reason, the research community spent a meaningful effort to develop automated methods to carry out this task. Data-driven and deep learning models have achieved a meaningful performance on the volumetric segmentation of the white matter bundles [9,10,

17,20]. A key factor for this result is the availability of large open datasets with annotated bundles [11,17].

The lack of analogous large open clinical datasets is preventing the training of supervised learning models for patients, where the anatomy deviates from the normative model. Despite the occasional use of the models trained on healthy individuals for some small collection of patients [3,12,18], the translation of these models to clinical data remains an open challenge.

The source of complexity is represented by the deviation of patients' data from the distribution of healthy population, and consequently the mismatch between training and inference data. Such a kind of mismatch is known in the literature as domain shift. The factors that contribute to domain shift are manifolds: the tumor's presence, the different operators segmenting the bundles, the methods of DWI acquisition and processing, and the quality of data. Transfer learning is the conventional approach for addressing domain shift [6].

In this study, we address the following research questions: i) what is the change in white matter bundle segmentation performance when a model trained in healthy individuals is applied in a group of glioma patients?; ii) is there a significant improvement in the segmentation performance in patients, when a small subset of the clinical data is used to adapt the model with transfer learning techniques?; and iii) can we characterize the portion of the recovered domain shift?

We focus our preliminary analysis on the corticospinal tract (CST) and on the inferior longitudinal fasciculus (ILF), two relevant tracts for preoperative planning of glioma resection.

The learning of the normative model is carried out by referring to an open dataset of hundreds of healthy individuals with annotations manually curated of the bundles [11]. The testing, the retraining, and the transfer learning are performed on a collection of 21 patients with glioma [21].

The results show that the drop in translating a trained model from healthy individuals to patients is 20% for the CST and 25% for ILF. Transfer learning is effective in recovering half of this gap by using a small set of glioma patients. The portion of domain shift related to systemic bias was indeed successfully recovered. On the other hand, the portion of errors due to alterations induced by the tumors is still an open challenge.

2 Materials

The data-driven investigation is designed upon two datasets: an open dataset of healthy individuals and a dataset of patients with gliomas. For each individual, we refer to the T1w brain images and the annotation of left CST and left ILF volumetric masks.

Data of healthy individuals are drawn from TractoInferno [11], a multisite collection of hundreds of MR acquisitions, both T1w and DWI, and their derivatives. Images were acquired with 3T scans with parameters detailed in [11].

Poulin et al. [11] distributed the dataset already preprocessed by performing for DWI brain extraction, de-noising, correction for eddy current, and N4 bias field, intensity normalization, and resampling; For T1w data Pouline et al. [11] performed denoising, N4 bias field correction, and registration. All images were manually checked for artifacts, and removed in cases of a poor-quality acquisition. The authors fitted the constrained spherical deconvolution (CSD) diffusivity model and performed ensemble tracking, with the deterministic [2], probabilistic [16], particle-filtered [5], and surface-enhanced [14] tracking algorithms [11]. Notably, annotations manually curated of several main white matter bundles are available. The bundle annotations were performed with RecoBundlesX [4,13], and then manually revised by experts to exclude bundles not reconstructed accurately. For this reason, the number of available annotations varies for each bundle, as reported in Table 1.

Data of patients with glioma are acquired in the S. Chiara Hospital of Trento with a 1.5 T scanner. Details on the acquisition are described in [21]. The preprocessing of the DWI comprised artifacts' correction for DWI thermal noise, Gibbs ringing, eddy current, motion distortion, and susceptibility-induced EPI distortion. We selected 21 patients with tumors located in the left hemisphere. For each patient we consider the following processing: (i) the manual segmentation of tumor-volume-mask onto the T1w image, (ii) the CSD diffusion model reconstruction, (iii) the probabilistic tractography, (iv) the manual segmentation of the left CST and ILF and the related volumetric masks. All data are registered to MNI space with a linear transformation. Figure 1 reports the tumors' distances from the two bundles in each patient.

The two datasets encode a twofold domain shift. The former is represented by the alteration of the bundles due to the presence of tumors and the consequent deviation from the normative model. The latter refers to a systemic bias composed of many factors such as the acquisition parameters, the tractography computation, and the manual annotation of bundles.

Table 1. Datasets statistics. Table reporting separately for the clinical and the healthy subjects dataset, the number of bundles considered in this study (columns: # Clinical and # Healthy), and mean and standard deviation bundle volumes expressed in cm^3 (columns: Vol. Clinical (cm^3) and Vol. Healthy (cm^3)).

Bundle	# Clinical	# Healthy	Vol. Clinical (cm^3)	Vol. Healthy (cm^3)
CST	21	241	71.2 ± 9.5	83.4 ± 13.4
	21	251	41.9 ± 10.	69.2 ± 18.2

3 Methods

3.1 The Supervised Learning Models

Although there are many different automatic models for white matter segmentation in literature, we focus our analysis on methods based on volumetric rep-

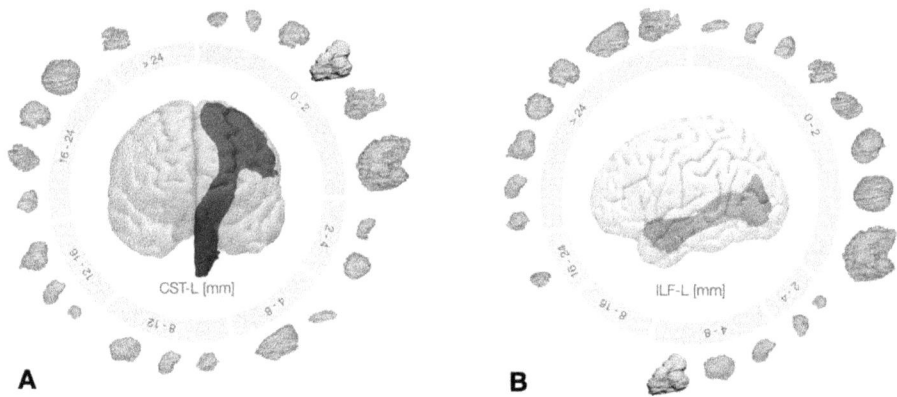

Fig. 1. Clinical dataset: (A) distance distribution in mm between the tumors and the left CST; (B) distance distribution in mm between the tumors and the left ILF

resentation. This choice aims to reduce the variance of bundle representation derived from tractography. The reference method for volumetric representation of bundle segmentation is a U-Net deep learning model [17]. The choice of the input to the model can span from the scalar intensity of T1w images to different DWI data derivatives such as fractional anisotropy (FA) or the main peaks of fiber orientation distribution function (fODF). A recent work [20] shows that T1w can be effective in capturing the spatial distribution of bundles and is a viable option when DWI data might be not available, such as in clinics. In this work, a patch-wise 3D U-net model is used. For these reasons, we design our analysis considering T1w images as reference data for bundle segmentation.

For our investigation, we consider the 3D U-Net as the reference learning model but with three different training strategies. The first strategy is designed as the default training process where the weights of the model are randomly initialized (*M-train*). The second strategy resumes the previous training process with a new set of data and, in the following, is referred to as retrain (*M-retrain*). Network weights are initialized by the last epoch of M-train and further optimized with the new training data. The third strategy is transfer learning [15] (*M-transfer*). It is analogous to the retraining strategy but with a different schema of weights' update. Learning is restricted to only the decoder of the 3D U-Net architecture, in charge of reconstructing the original space from latent space [15].

3.2 Performance Evaluation

To quantify the performance of the learning models after the different training strategies we use two measures. The first is the Dice Score Coefficient (DSC) that evaluates voxel-wise the similarity between the predicted bundle mask and the related ground truth defined by manual annotation. The second metric aims to support a pairwise comparison of two training strategies at the voxel level when we operate the inference on the same collection of data. The goal is to evaluate the spatial distribution of the regions where one training strategy outperforms the other. In the following, we refer to this second measure as *Improvement Map*.

The computation of the Improvement Map takes as input a set of brain images of different individuals, linearly registered to MNI. For each of them, two bundle masks are generated by two distinct models as predictions of bundle segmentation. Each of these bundle masks is converted into a binary true prediction map. This binary map has values 1 in voxels where either negative or positive true predictions occur, 0 otherwise. An individual Improvement Map is obtained by computing the difference between the true prediction maps derived by the two models applied to the same image. When the behavior of the two models is the same, the voxels have a value of 0, otherwise 1, or -1. We may obtain a global Improvement Map by averaging voxel-wise the individual maps, across the population. Given two competing models, positive values of the Improvement Map account for better performance of the former model, and negative values for the latter model. Values close to 0 denote similar behavior.

3.3 Domain Shift Assessment

The characterization of the domain shift between the healthy population and the glioma patients population is carried out following two perspectives: one oriented to the volumetric representation and another to the tractography representation. The first approach measures the spatial distribution of the bundle masks with two probabilistic maps, one for healthy individuals, and the other for glioma patients. The second metric, tractography-based, is measured along the main pathway of a bundle, represented with the most representative fiber. This fiber is sometimes referred to as the *skeleton* [8] of the bundle. The intuitive idea is to measure the deformation of a bundle with respect to the bundle-normative model along the skeleton with registration. For this purpose, we introduce the notion of *Warp V-Norm* and *Warp V-Norm profile*.

The *Warp V-Norm* estimates the local shape differences between two bundles, encoded as volumetric binary masks after a spatial normalization with an affine registration in a common space. The first step is the computation of the warp by registering the two bundles, combining a rigid and a diffeomorphic transformation. The Warp V-Norm is the 2-norm of the vector encoding the whole transformation in each voxel of the warped image. High values denote a major alteration of a given bundle with respect to the normative model.

The *Warp V-Norm profile* summarizes the volumetric information of deformation with respect to the skeleton of the normative model. The reference skeleton is derived from the fiber representation of all the bundles of the healthy population dataset and represents the average pathway in the maximum density area of the streamlines [1]. The Warp V-Norm profile is a vector where each point of the fiber is associated with the Warp V-Norm of the correspondent voxel.

4 Experiments and Results

The experiments aim to assess the performance of the supervised learning models trained according to the three reference strategies: no transfer learning, model retraining, and transfer learning. The additional goal is to characterize the behavior of the different learning strategies with respect to the different factors of domain shift between healthy and patients. All three training sessions and relative performance assessments were carried out for both bundles. For CST we characterized also the domain shift, and how M-retrain and M-transfer can handle it.

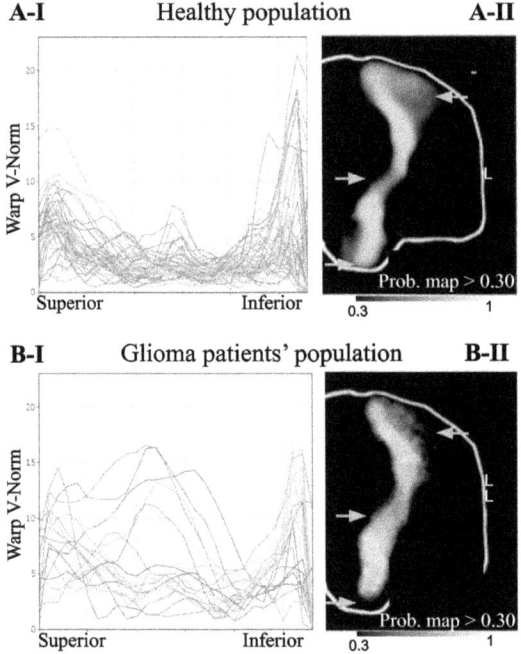

Fig. 2. Normative CST and clinical deviations: (A-I): Warp V-Norm profile for healthy test set with respect to normative model along the skeleton pathway; (A-II): Probability map distribution thresholded at 0.3 in healthy population; (B-I) and (B-II): similar to previous panels but for patients. The arrows indicate where the probability distributions differ more between healthy and patient sets

We first focus on how to measure the domain shift between the two datasets for CST, considering the volumetric representation of the bundles as voxel masks. Tractography representation of bundles is first mapped to voxels, then smoothed and registered with an affine transformation to the MNI common reference space. Separately for the two datasets, the probability maps measuring the spatial distribution of the bundles are computed as the relative frequency of each voxel to be part of the CST. To highlight the difference between the two maps we provide a thresholded version ($probability > 0.30$) in Figs. 2.A-II and 2.B-II.

For characterizing the CST domain shift with respect to the tractography representation, we compute the Warp V-Norm profile for the subjects in the test set of the healthy population and for the 21 glioma patients. The template of bundle mask for the normative model is derived from the probability map of healthy individuals, thresholded so that its volume matches the mean volume of the whole population. Rigid and diffeomorphic registrations of each individual's bundle with respect to the template are computed using ANTs. The displacement along the skeleton for each individual with respect to the normative model is separately reported in Fig. 2.A-I and 2.B-I, healthy and patients respectively.

Then we trained the bundle segmentation models for both bundles. To reduce the computational overload and to balance the ratio between the classes of the binary classifiers we resize the bounding box of T1w images, by excluding the contralateral hemisphere, and other regions where the bundles can't anatomically be located.

As reference implementation of 3D U-Net, we refer to nnU-Net [7], a state-of-the-art deep-learning framework for biomedical image segmentation. Intensity normalization is included in the default setting of the nnU-Net that performs z-score intensity normalization.

We design 3 training sessions carried out separately for each bundle, for a total of 6 trained models. For each bundle, in the first session, a 3D U-Net model is trained with healthy data (M-train), in the second the M-train model is retrained with patients' data only (M-retrain), in the third the M-train model is trained with a transfer learning method with patients' data only (M-transfer). Each of these three models is used for inference both on healthy and patient test sets. Performance is measured as the average DSC between predicted and true bundle masks. Results are reported in Table 2.

Train and test split is designed differently for healthy and patients data. The former is divided with a ratio of 80/20. The learning is carried out with a three fold cross-validation schema and 50 epochs each. Differently, for patients, where the sample size is small, we follow a stratified cross-validation schema. Data are partitioned according to a 5-fold cross-validation, and a nested 3-fold partition is performed with 200 epochs each. The variance of bundles is much larger in patients. For this reason, the design of partitioning is revised to balance the distribution of bundles with major alterations. As a measure of alteration, we refer to the Warp V-Norm profile computed for both bundles as reported above.

For the CST bundle, we carried out a pairwise comparison of models' performance at the voxel level, by computing the Improvement Maps, looking at the True Positive and True Negative errors, as illustrated above. We focus our analysis on the contrast between M-transfer and M-train models when inference is operated on patients. The improvement map of True Positive and True Negative reports the spatial distribution where M-transfer outperforms M-train. Figure 3 shows in red the portion of domain shift successfully managed by the transfer learning.

Table 2. Learning assessment. Mean DSC and standard deviations of the inference on healthy and patient datasets.

Bundle	Model	Healthy dataset	Patients dataset
CST	M-train (Healthy)	0.82 ± 0.02	0.66 ± 0.06
CST	M-retrain (Patient)	0.67 ± 0.03	0.74 ± 0.09
CST	M-transfer (Patient)	0.75 ± 0.02	0.74 ± 0.08
ILF	M-train (Healthy)	0.79 ± 0.05	0.59 ± 0.09
ILF	M-retrain (Patient)	0.59 ± 0.05	0.68 ± 0.10
ILF	M-transfer (Patient)	0.67 ± 0.04	0.67 ± 0.10

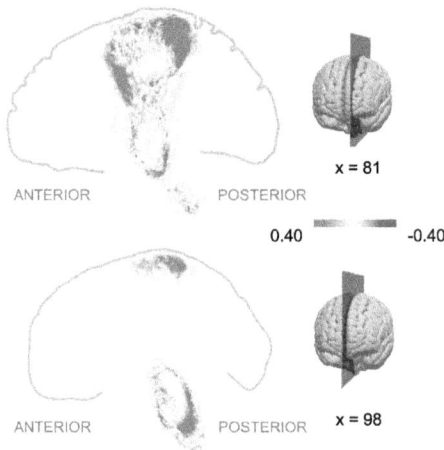

Fig. 3. Improvement Maps of CST. A voxelwise analysis to compare M-train and M-transfer models when used for inference on patients' data. The red color highlights the region where M-transfer outperforms M-train, the blue color otherwise. (Color figure online)

5 Discussion and Conclusions

Results in Table 2 report the meaningful drop in performance when the M-train model is not applied to the healthy population, but to patients' data: for CST the DSC dropped from 0.82 to 0.66, while for ILF from 0.79 to 0.59.

The reason for this loss in accuracy is the domain shift, depicted in Fig. 2 for CST, and also represented in the difference in mean volume for ILF as reported in Table 1.

The probability maps of CST differ between two populations in the inferior, middle, and superior portions of the bundles, as shown by Figs. 2.A-II and 2.B-II. The differences in the inferior and in the superior parts seem related to systemic bias of patients' data: (i) lack of inferior slices in MR acquisition for patients, and (ii) tractography reconstruction in the superior part of CST. Since there are no systemic biases that may cause alterations in the middle part of the bundle, the shift in this region is probably caused by tumor-induced alterations. The tumor-related domain shift in the middle of the bundle is also represented by the Warp V-Norm profiles in Fig. 2.B-I.

Both M-retrain and M-transfer are effective in partially recovering the domain shift on glioma patients: DSC performance of M-retrain and M-transfer were both significantly higher than the ones of M-train for both bundles (p. value < 0.01 of the Wilcoxon signed-rank test). For CST, in the clinical dataset, we observe a recovery of DSC performance from 0.66 with M-train to 0.74 for both M-retrain and M-transfer. For ILF, in the clinical dataset, we observe a recovery of DSC performance from 0.59 for M-train, to 0.68 and 0.67 for M-retrain and M-transfer. Remarkably, the models capture the different distribution of patients' data, with only 10% of the data used for training the original normative model.

While inference on patients' data is substantially equivalent for M-retrain and M-transfer, there is a meaningful difference if we focus on the inference on the healthy population, with M-transfer that performs significantly better than M-retrain for both bundles (p.value < 0.01 of the Wilcoxon signed-rank test). As expected, M-transfer better preserves the balance between the two populations.

For CST we carried out an analysis to understand which domain shift portion was recovered by transfer learning. With the analysis of recovery in Fig. 3 we localized the regions of CST where the accuracy increases after the transfer learning. These regions are located in the inferior part of the stem and in the posterior part of the parietal terminations of CST. As pointed out before, these differences located in the superior and in the inferior parts of the CST seem related to systemic bias of patients' data. Apparently, there is no improvement in the middle part where deviations from the normative model due to tumors are located.

In conclusion, both transfer learning and retraining can exploit efficiently a small set of clinical data to recover part of the domain shift bias. Transfer learning better preserves the information of bundles in the healthy population. We contribute to the understanding of how domain shift is affecting bundle segmentation in glioma patients. This work provides the basis for future developments to improve the management of domain shift due to tumor bias, for example extending the analysis to other bundles and to larger clinical datasets.

Compliance with Ethical Standards. This study was conducted following the ethical standards of the Declaration of Helsinki and was approved by the local ethical committee (authorization ID A734).

Disclosure of Interests. This work was partially supported by the grant PAT Reg. n. 764/2021 NeuSurPlan. We also acknowledge the support of the PNRR project FAIR - Future AI Research (PE00000013), under the NRRP MUR program funded by the NextGenerationEU.

References

1. Amorosino, G., Olivetti, E., Jovicich, J., Avesani, P.: How does white matter registration affect tractography alignment? In: 2023 IEEE 20th International Symposium on Biomedical Imaging (ISBI), pp. 1–5 (2023). https://doi.org/10.1109/ISBI53787.2023.10230615, https://ieeexplore.ieee.org/document/10230615, iSSN: 1945-8452
2. Basser, P.J., Pajevic, S., Pierpaoli, C., Duda, J., Aldroubi, A.: In vivo fiber tractography using DT-MRI data. Magn. Reson. Med. **44**(4), 625–632 (2000). https://doi.org/10.1002/1522-2594(200010)44:4<625::aid-mrm17>3.0.co;2-o
3. Bertò, G., et al.: Classifyber, a robust streamline-based linear classifier for white matter bundle segmentation. NeuroImage **224**, 117402 (2021). https://doi.org/10.1016/j.neuroimage.2020.117402, https://www.sciencedirect.com/science/article/pii/S1053811920308879
4. Garyfallidis, E., et al.: Recognition of white matter bundles using local and global streamline-based registration and clustering. NeuroImage **170**, 283–295 (2018). https://doi.org/10.1016/j.neuroimage.2017.07.015, https://linkinghub.elsevier.com/retrieve/pii/S1053811917305839
5. Girard, G., Whittingstall, K., Deriche, R., Descoteaux, M.: Towards quantitative connectivity analysis: reducing tractography biases. NeuroImage **98**, 266–278 (2014).https://doi.org/10.1016/j.neuroimage.2014.04.074, https://www.sciencedirect.com/science/article/pii/S1053811914003541
6. Hosna, A., Merry, E., Gyalmo, J., Alom, Z., Aung, Z., Azim, M.A.: Transfer learning: a friendly introduction. J. Big Data **9**(1), 102 (2022). https://doi.org/10.1186/s40537-022-00652-w
7. Isensee, F., Jaeger, P.F., Kohl, S.A.A., Petersen, J., Maier-Hein, K.H.: nnU-Net: a self-configuring method for deep learning-based biomedical image segmentation. Nat Methods **18**(2), 203–211 (2021). https://doi.org/10.1038/s41592-020-01008-z, https://www.nature.com/articles/s41592-020-01008-z Publisher: Nature Publishing Group
8. Li, W., Hu, X.: Robust tract skeleton extraction of cingulum based on active contour model from diffusion tensor MR imaging. PLoS ONE **8**(2), e56113 (2013). https://doi.org/10.1371/journal.pone.0056113, https://dx.plos.org/10.1371/journal.pone.0056113
9. Lu, Q., Li, Y., Ye, C.: Volumetric white matter tract segmentation with nested self-supervised learning using sequential pretext tasks. Medical Image Anal. **72**, 102094 (2021).https://doi.org/10.1016/j.media.2021.102094, https://www.sciencedirect.com/science/article/pii/S1361841521001407
10. Nelkenbaum, I., Tsarfaty, G., Kiryati, N., Konen, E., Mayer, A.: Automatic segmentation of white matter tracts using multiple brain MRI sequences. In: 2020 IEEE 17th International Symposium on Biomedical Imaging (ISBI), pp. 368–371 (2020). https://doi.org/10.1109/ISBI45749.2020.9098454, iSSN: 1945-8452
11. Poulin, P., et al.: TractoInferno - a large-scale, open-source, multi-site database for machine learning dMRI tractography. Sci. Data **9**(1), 725 (2022). https://doi.org/10.1038/s41597-022-01833-1, https://www.nature.com/articles/s41597-022-01833-1 Publisher: Nature Publishing Group
12. Richards, T.J., Anderson, K.L., Anderson, J.S.: Fully automated segmentation of the corticospinal tract using the TractSeg algorithm in patients with brain tumors. Clin. Neurol. Neurosurgery **210**, 107001 (2021). https://doi.org/10.1016/j.clineuro.2021.107001, https://www.sciencedirect.com/science/article/pii/S0303846721005308

13. St-Onge, E.: Analyse de l'architecture de la matière blanche et projection de mesures sur la surface corticale. Ph.D. thesis (2021). https://savoirs.usherbrooke.ca/handle/11143/17888?locale-attribute=fr
14. St-Onge, E., Daducci, A., Girard, G., Descoteaux, M.: Surface-enhanced tractography (SET). NeuroImage **169**, 524–539 (2018). https://doi.org/10.1016/j.neuroimage.2017.12.036, https://www.sciencedirect.com/science/article/pii/S1053811917310583
15. Sundaresan, V., Zamboni, G., Dinsdale, N.K., Rothwell, P.M., Griffanti, L., Jenkinson, M.: Comparison of domain adaptation techniques for white matter hyperintensity segmentation in brain MR images. Medical Image Anal. **74**, 102215 (2021). https://doi.org/10.1016/j.media.2021.102215, https://www.sciencedirect.com/science/article/pii/S1361841521002607
16. Tournier, J.D., Calamante, F., Connelly, A.: MRtrix: diffusion tractography in crossing fiber regions. Int. J. Imaging Syst. Technol. **22**(1), 53–66 (2012). https://doi.org/10.1002/ima.22005, https://onlinelibrary.wiley.com/doi/abs/10.1002/ima.22005, eprint: https://onlinelibrary.wiley.com/doi/pdf/10.1002/ima.22005
17. Wasserthal, J., Neher, P., Maier-Hein, K.H.: TractSeg - Fast and accurate white matter tract segmentation. NeuroImage **183**, 239–253 (2018). https://doi.org/10.1016/j.neuroimage.2018.07.070, https://www.sciencedirect.com/science/article/pii/S1053811918306864
18. Xue, T., et al.: Superficial white matter analysis: an efficient point-cloud-based deep learning framework with supervised contrastive learning for consistent tractography parcellation across populations and dMRI acquisitions. Med. Image Anal. **85**, 102759 (2023). https://doi.org/10.1016/j.media.2023.102759, https://www.sciencedirect.com/science/article/pii/S1361841523000208
19. Yang, J., Yeh, C.H., Poupon, C., Calamante, F.: Diffusion MRI tractography for neurosurgery: the basics, current state, technical reliability and challenges. Phys. Med. Biol. **66**(15), 15TR01 (2021). https://doi.org/10.1088/1361-6560/ac0d90, https://dx.doi.org/10.1088/1361-6560/ac0d90, publisher: IOP Publishing
20. Yang, Q., et al.: Learning white matter subject-specific segmentation from structural MRI. Med. Phys. **49**(4), 2502–2513 (2022). https://doi.org/10.1002/mp.15495
21. Zigiotto, L., et al.: Segregated circuits for phonemic and semantic fluency: a novel patient-tailored disconnection study. NeuroImage: Clin. **36**, 103149 (2022). https://doi.org/10.1016/j.nicl.2022.103149, https://www.sciencedirect.com/science/article/pii/S2213158222002145

Image Quality Transfer of Diffusion MRI Guided By High-Resolution Structural MRI

Alp G. Cicimen[1(✉)], Henry F. J. Tregidgo[1], Matteo Figini[1],
Eirini Messaritaki[2], Carolyn B. McNabb[2], Marco Palombo[2],
C. John Evans[2], Mara Cercignani[2], Derek K. Jones[2],
and Daniel C. Alexander[1]

[1] Centre for Medical Image Computing, University College London, London, UK
{gundogan.cicimen.22,h.tregidgo,m.figini,d.alexander}@ucl.ac.uk
[2] Cardiff University Brain Research Imaging Centre, Cardiff University, Cardiff, UK
{MessartiakiE2,mcnabbc1,palombom,evansj31,cercignanim,
JonesD27}@cardiff.ac.uk

Abstract. Prior work on the Image Quality Transfer on Diffusion MRI (dMRI) has shown significant improvement over traditional interpolation methods. However, the difficulty in obtaining ultra-high resolution Diffusion MRI scans poses a problem in training neural networks to obtain high-resolution dMRI scans. Here we hypothesise that the inclusion of structural MRI images, which can be acquired at much higher resolutions, can be used as a guide to obtaining a more accurate high-resolution dMRI output. To test our hypothesis, we have constructed a novel framework that incorporates structural MRI scans together with dMRI to obtain high-resolution dMRI scans. We set up tests which evaluate the validity of our claim through various configurations and compare the performance of our approach against a unimodal approach. Our results show that the inclusion of structural MRI scans do lead to an improvement in high-resolution image prediction when T1w data is incorporated into the model input.

Keywords: Super-Resolution · Multi-modality · Diffusion MRI

1 Introduction

Image Quality Transfer (IQT) [1] is a machine learning framework that estimates how MRI scans should look if acquired on state-of-the-art scanners. The first, random forest, implementation showed promise in upsampling diffusion MRI (dMRI) images and has been further developed using both conventional convolutional neural network (CNN) approaches to upsample dMRI scans [20] and other types of MRI [14] and diffusion network models to upsample low-field structural MRI [12]. While these works do demonstrate the capability of super-resolution models to achieve realistic results, their upsampling capabilities are limited by their training configurations. This problem is not easily rectifiable on models that only use dMRI scans as input as obtaining a sufficient Signal-to-Noise Ratio (SNR) at high resolution is challenging, especially for dMRI models which require several volumes to be acquired [11].

Therefore, we propose the incorporation of a secondary MRI modality in conjunction with the low-resolution dMRI input to mitigate the outlined limitations of models that use a single data modality. Prior works have demonstrated that a secondary structural MRI modality of different contrast can be incorporated into a model to achieve state-of-the-art results in upsampling structural MRI data [15]. We hypothesise that this incorporation of a second modality can also be used in improving the upsampling capabilities of dMRI In addition, structural MRI scans are easier to obtain at ultra-high resolution and are more readily available, which can help with bypassing the limitations of the dMRI training data. Our approach aims to show as a proof-of-concept that structural MRI scans contain inherent information that a CNN can optimize for and therefore generate better high-resolution dMRI-based computational models.

2 Methods

The inputs to our model are a low-resolution dMRI volume and a high-resolution T1w scan. We use a similar framework to SynthSR [9] in that inputs of varying resolutions are upsampled to a specified target resolution. However, instead of processing a whole image volume we use a patch-based network to reduce memory pressure. In addition, instead of upsampling specifically to 1 mm isotropic resolutions, we implicitly incorporate the desired target resolution in the form of structural information from the T1w companion volume. We explain our network architecture and reasoning further in Subsect. 2.1.

For the dMRI-based neural network input, we use the Diffusion Tensor Imaging (DTI) model [2]. We specifically use 6 independent elements of the diffusion tensor as inputs. In Subsect. 2.2 and Subsect. 2.3 we explain how we process our input data, and how our training and evaluation configurations differ from each other.

2.1 Network Architecture

For this network we use a multimodal neural network based on a 3D-UNet architecture [6] which has shown versatility and capability in prior super-resolution works [9,14]. Due to previous works outlining better performance at capturing common features [5,16,19], we fuse information from our dMRI and T1w images at the latent layer. Our modified architecture uses a secondary encoder for the T1w input and joins the two latent encodings together using an averaging operation. For each encoder and decoder layer, we have repeating convolutions without residual connections to further capture possible information. We include skip connections between the decoder layers and layers from both encoders at the corresponding level, allowing transfer of high-level information from both modalities. A visual diagram of our model can also be found in Fig. 1.

Our model inputs are a high-resolution T1w patch $\mathcal{P}^s_{HR}$ of shape ($16 \times 16 \times 16 \times 1$) and a DTI low-resolution patch $\mathcal{P}^d_{LR}$ of shape ($16 \times 16 \times 16 \times 6$) and outputs a high-resolution DTI patch $\mathcal{P}^d_{HR}$ of shape ($16 \times 16 \times 16 \times 6$). Here we use a pre-upsampling approach, where the input dMRI image is initially linearly upsampled to the resolution of patch $\mathcal{P}^s_{HR}$ to generate patch $\mathcal{P}^d_{LR}$ before both are passed into the model. This allows a dynamic upsampling rate to the T1w resolution which would not be possible with a post-upsampling model, where the model has a preconfigured upsampling rate.

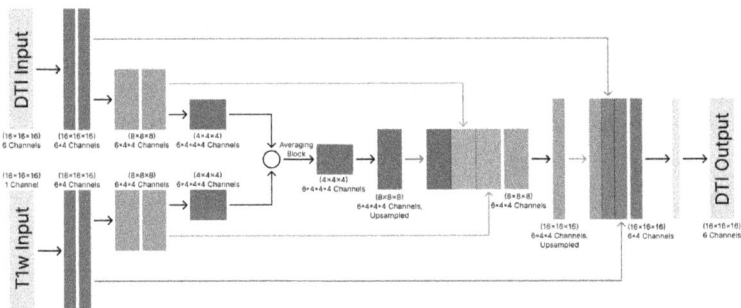

Fig. 1. Network diagram for our multimodal UNet architecture designed to test our multimodal approach to super-resolution. For this initial feasibility test we limit the network to three levels.

In addition to our fusion model, we use a unimodal UNet for comparison in ablation experiments. This is a standard, single-encoder UNet architecture that only uses the DTI as input. We have configured our base model such that the number of hidden layers and channels allocated to the DTI encoder is the same as the channels allocated in the DTI only ablation encoder. This means any increase in performance can be attributed to the T1w weights rather than an increase in available weights for the DTI input.

2.2 Training Configuration

To train our model we need to supply triplets of patches, $\mathcal{P}^D_{HR}$, $\mathcal{P}^D_{LR}$ and $\mathcal{P}^S_{HR}$, ensuring these three patches are sampled on the same discretisation. As the T1w images in our training set are higher resolution than the available high-resolution DTI, we need to downsample our T1w images while correcting for aliasing effects. For this, we use the same approach as SynthSeg [3,4] to smooth the input T1w with an appropriate Gaussian kernel for the target resolution before downsampling.

Our input low resolution DTI is generated using a similar process, starting by Gaussian blurring before downsampling with trilinear interpolation to our low-resolution space from the original DTI data. We then linearly upsample the low-resolution DTI to the target discretisation of our T1w input. This results in images with voxel spacing at the target resolution, where the voxels are the same shape as our target image, but have a lower effective resolution. For our base model we randomly select one of four effective $\mathcal{P}_{LR}^D$ resolutions to use as our input configuration at each iteration during training. This is done to ensure that the model does not adapt to a specific upsampling rate, and to provide further robustness to our model. Our training and testing resolution configurations can be found in detail in Sect. 3.

After the resampling steps, we clip our data to a specific range and apply min-max normalisation to both the DTI and T1w inputs. For DTI we clip and normalise our diagonal elements D_{xx}, D_{yy} and D_{zz} to range $[0, 2 \times 10^{-3}]$ mm^{2}s^{-1}, and our non-diagonal elements D_{xy}, D_{xz} and D_{yz} to $[-2 \times 10^{-3}, 2 \times 10^{-3}]$ mm^{2}s^{-1}. We have found that these clipping values preserve the information in the non-CSF regions, while dynamic clipping can excessively compress values in these regions. For the T1w input we clip and normalise our image to between the 2nd and 98th percentile of masked voxel intensities in our training set.

The HR DTI, HR T1w and LR DTI scans for each training subject are then split into patches of size $16 \times 16 \times 16$ which we concatenate into a patch triplet of shape (16, 16, 16, 13). During training, we add random augmentations in the form of noise, brightness adjustments and gamma scaling into the T1w input to ensure robustness of the model.

2.3 Evaluation and Inference Configuration

To evaluate our model we use similar preprocessing steps as in training to generate input patch pairs. However, patches are selected such that neighbouring patches overlap by four voxels in each dimension. To recombine patches into a full volume we blend patch edges by averaging the overlapping regions resulting in a better image output quality.

After the model outputs are reconstructed into a single image, we scale back the output images to the original DTI scale using the clipped parameters outlined in Subsect. 2.2. The resulting data can now be evaluated as is, or with DTI-based metrics as explained in Sect. 3.

3 Experiments

This section presents the experimentation procedure that we used to evaluate our model. For each setup we tested in isolation a specific aspect of our model against an ablated one, which was configured as either a unimodal network or a model that used a single training input resolution.

3.1 Experiment Data

We use scans from two different data sources for our quantitative and qualitative experiments. For our quantitative experiments, we use the datasets in the WU-Minn consortium from the Human Connectome Project [21] for training and testing purposes. The consortium is a multi-institutional collaboration that contains multiple datasets consisting of MRI and DWI sequences obtained using a customized Siemens 3T scanner with a gradient field strength of 300 mT/m. We use 174 subjects from the HCP Young Adult dataset, from which we specifically use the T1w scan with 0.7 mm isotropic voxel spacing, as well as a DWI sequence that consists of 90 diffusion-weighted images with b-values of $1000 \, s/mm^2$, with a voxel size of 1.25 mm. The sequence contains 18 $b = 0$ scans as well, resulting in a total of 108 dMRI scans and 1 T1w image per subject that we use. We calculate our model inputs by fitting the DTI model to the dMRI images. We split our data on a per-subject base into three parts, which act as our training, validation and testing sets with a split percentage of 70%, 10% and 20% respectively.

For our qualitative results, we use a second data source for evaluating the network that has been trained on the HCP dataset. For this purpose, we use an out-of-distribution scan obtained through two different MRI scanners. Our DWI sequence was obtained through a Connectom 3T scanner using gradient field strengths of 300 mT/m, and our structural scan was obtained through a 7T scanner. While the specifications of the Connectom scanner are the same as the scanner used for obtaining the dMRI scans in the HCP dataset, our DWI sequence is limited in terms of image resolution of 2 mm isotropic voxel size, and it consists of 53 images with a maximum b-value of $b = 1200 \, s/mm^2$. We register our T1w to the linearly interpolated 0.7 mm DTI image using EasyReg [8].

3.2 Experiment Setup

To evaluate the model performance, we have devised 4 tests, with the configurations described in Table 1, that each individually evaluate in-distribution performance, performance on unseen upsampling factors, and performance when targeting unseen higher resolutions. The 4 tests were devised as follows:

Test 1: Baseline. Our initial test compares the baseline performance of the multimodal approach to three ablation models. For this test, we train our network on multiple low-resolution DTI inputs to upsample to our target resolution. The model is then evaluated on 2 in-distribution upsampling rates, a 1.5× upsampling rate (from 1.875 to 1.25 mm voxel dimensions) and a 2.5× upsampling rate (from 3.125 to 1.25 mm voxel dimensions). This tests the general performance of the multi-resolution training against corresponding single-resolution networks. In addition, we compare the model's performance to an ablation model that uses DTI input but no T1w input.

Test 2: Out of Distribution Upsampling. Here we test our models on upsampling an out-of-distribution 3.75 mm DTI image input to evaluate how the models in Test 1 behave on previously unseen upsampling rates. This tests if the

Table 1. Description of network parameters for the four experiments. Here we list the isotropic input and target voxel dimensions used in training and testing. Grouped resolutions indicate the network was trained or tested on multiple upsampling rates.

Test Name	Model Configuration	T1w Included?	Training DTI Input Resolution(s) (mm)	Training Target Resolution (mm)	Testing DTI Input Resolution(s) (mm)	Testing Target Resolution (mm)
1: Baseline	Standard	Yes	{1.5625, 1.875, 2.5, 3.125}	1.25	{1.875, 3.125}	1.25
	Single Resolution 1.5625 mm	Yes	1.5625			
	Single Resolution 3.125 mm	Yes	3.125			
	Multiple Resolution DTI only	No	{1.5625, 1.875, 2.5, 3.125}			
2: OOD Upsampling Rate	Standard	Yes	{1.5625, 1.875, 2.5, 3.125}	1.25	3.75	1.25
	Single Resolution 1.5625 mm	Yes	1.5625			
	Single Resolution 3.125 mm	Yes	3.125			
	Multiple Resolution DTI only	No	{1.5625, 1.875, 2.5, 3.125}			
3: OOD Target Resolution	Standard	Yes	{2.8, 3.0, 3.2, 3.4}	2.0	2.0	1.25
	Single Upsample Rate	Yes	3.2			
	Single Upsample Rate DTI only	No	3.2			
4: Qualitative	Standard	Yes	{1.5625, 1.875, 2.5, 3.125}	1.25	2.0	0.7
	Single Upsample Rate DTI only	No	3.75			

addition of the T1w image compensates for the lost details in further down-sampled DTI image. A higher performance of the multimodal network over our unimodal ablation model would demonstrate that the T1w image contains useful information for upsampling DTI data.

Test 3: Out of Distribution Target Resolution. For our final quantitative test we evaluate whether we can achieve resolutions higher than those available in the training set. We train our network on upsampling DTI images of multiple resolutions to a single target resolution $\mathcal{R}'_{train}$, selected such that $\mathcal{R}'_{train}$ is worse than the test resolution $\mathcal{R}'_{test}$. For this test, our ablation models differ by using only a single upsampling rate $\mathcal{U}_{ablation} = \mathcal{R}'_{train}/\mathcal{R}'_{test}$. This is because there isn't an explicit method to use a dynamic upsampling rate without the use of extra parameters. Here an improved performance of the multimodal network over the unimodal model would demonstrate its ability to infer information from the T1w image and use it to correct features not discernible from lower quality DTI input.

Test 4: Qualitative. In addition to the quantitative tests we also include a result for the qualitative output of our model compared to an unimodal ablation model trained on a 3× upsampling rate. We use the Coloured FA map to visualise how the addition of T1w input from a 7T scanner affects the high-resolution model output.

To obtain our quantitative results, we used 34 test subjects from the HCP dataset that were selected at random. We evaluate the performance of each model using the metrics described in Subsect. 3.3 and evaluate the statistical significance of differences using the Wilcoxon's Signed Rank test with a significance threshold of 5%. Our qualitative result was obtained from a single test subject acquired as outlined in Subsect. 3.1.

We have trained our models for 100 epochs, with 400 patches selected per subject per epoch. We use a minibatch size of 40 per iteration and use the L1 loss through the batch as our loss function. For optimisation, we have used Adam [13] and an initial learning rate of 1×10^{-3}, with an exponential learning rate decay that halves every 10 epochs.

3.3 Experiment Evaluation

To evaluate our model performance we have used both the DTI output directly and its derived metrics. We initially evaluate the model performance on the DT-RMSE metric [1] as

$$\text{median}_{v \in \Omega(i)} \left(\sqrt{\frac{1}{6} \sum_{j=1}^{6} (D_j - D_j^*)^2} \right) \quad (1)$$

where D_j and D_j^* are the predicted and ground truth j-th DTI element for every voxel v that is contained within the set of all masked regions in a single subject $\Omega(i)$. For the derived metrics we consider the Mean Diffusivity (MD) and Fractional Anisotropy (FA) and evaluate the Root-Mean-Squared Error (RMSE) as well as the Structural Similarity Index Measure (SSIM) [24], which assesses the general visual performance of our model on the aforementioned metrics. In addition, we investigate the Coloured FA maps [18] using the mean cosine similarity (CSIM) between the predicted and target vectors, defined as

$$\text{CSIM}(\mathbf{x}, \mathbf{x}^*) = \frac{1}{|\Omega|} \sum_{i \in \Omega} \left(\frac{|\mathbf{x}_i^* \cdot \mathbf{x}_i|}{||\mathbf{x}_i^*|| \cdot ||\mathbf{x}_i||} \right) \quad (2)$$

4 Results

The quantitative results of our experiments can be viewed in Table 2.

Our baseline in-distribution results demonstrate that the incorporation of a T1w input improves the model's prediction capabilities. The difference between the result metrics of our multiple resolution approach and our unimodal ablation model was statistically significant, with the median subject-wise RMSE and similarity metrics favouring our multimodal approach for both FA and MD. However, for 1.875 mm input resolution we noticed that the single resolution ablation model trained at its native input resolution performed the best out of all inputs. We however have noticed that the aforementioned model tended to

Table 2. Median RMSE and SSIM values across test subjects for experiments one to three. For each experiment, the best performing model is underlined and highlighted in bold when differences to the second best model reach significance determined by a Wilcoxon's signed rank test.

Test Name	Model Configuration	DT - RMSE↓	MD RMSE↓	FA RMSE↓	MD SSIM↑	FA SSIM↑	CFA CSIM↑
1: Baseline 1.875 mm Input	Linear Interpolation	6.711×10^{-5}	1.463×10^{-4}	1.005×10^{-1}	0.962	0.943	0.930
	Multiple Resolution Standard Model	3.489×10^{-5}	6.060×10^{-5}	4.537×10^{-2}	0.993	0.986	0.963
	Single Resolution 1.875 mm	$\mathbf{\underline{3.337 \times 10^{-5}}}$	$\mathbf{\underline{5.794 \times 10^{-5}}}$	$\mathbf{\underline{4.504 \times 10^{-2}}}$	$\underline{0.994}$	$\underline{0.987}$	$\underline{0.964}$
	Single Resolution 3.125 mm	1.427×10^{-4}	1.725×10^{-4}	1.593×10^{-1}	0.965	0.944	0.939
	Multiple Resolution No T1w Input	3.568×10^{-5}	6.744×10^{-5}	4.767×10^{-2}	0.992	0.986	0.962
1: Baseline 3.125 mm Input	Linear Interpolation	1.028×10^{-4}	2.068×10^{-4}	1.504×10^{-1}	0.923	0.879	0.890
	Multiple Resolution Standard Model	4.701×10^{-5}	$\mathbf{\underline{8.725 \times 10^{-5}}}$	5.966×10^{-2}	$\mathbf{\underline{0.985}}$	$\underline{0.976}$	0.946
	Single Resolution 1.875 mm	7.803×10^{-5}	1.810×10^{-4}	1.091×10^{-1}	0.943	0.935	0.911
	Single Resolution 3.125 mm	$\mathbf{\underline{4.624 \times 10^{-5}}}$	9.474×10^{-5}	$\underline{5.938 \times 10^{-2}}$	0.983	0.974	$\underline{0.948}$
	Multiple Resolution No T1w Input	5.277×10^{-5}	1.170×10^{-4}	6.736×10^{-2}	0.977	0.968	0.940
2: Out of Dist. Upsampling Rate	Linear Interpolation	1.141×10^{-4}	2.259×10^{-4}	1.734×10^{-1}	0.912	0.857	0.871
	Multiple Resolution Standard Model	7.795×10^{-5}	1.557×10^{-4}	1.168×10^{-1}	0.955	0.931	0.911
	Single Resolution 1.875 mm	1.032×10^{-4}	2.147×10^{-4}	1.522×10^{-1}	0.921	0.888	0.882
	Single Resolution 3.125 mm	$\mathbf{\underline{5.802 \times 10^{-5}}}$	$\mathbf{\underline{1.195 \times 10^{-4}}}$	$\mathbf{\underline{8.015 \times 10^{-2}}}$	$\underline{0.974}$	$\underline{0.959}$	$\underline{0.932}$
	Multiple Resolution No T1w Input	8.617×10^{-5}	1.788×10^{-4}	1.265×10^{-1}	0.940	0.911	0.897
3: Out of Dist. Target Resolution	Linear Interpolation	7.222×10^{-5}	1.612×10^{-4}	1.053×10^{-1}	0.953	0.930	0.926
	Multiple Resolution Standard Model	5.280×10^{-5}	1.060×10^{-4}	7.506×10^{-2}	0.976	0.962	0.943
	With T1w Input 1.6× Upsampling Rate	$\mathbf{\underline{4.982 \times 10^{-5}}}$	$\mathbf{\underline{9.798 \times 10^{-5}}}$	$\mathbf{\underline{6.759 \times 10^{-2}}}$	$\underline{0.980}$	$\underline{0.968}$	$\underline{0.945}$
	Without T1w Input 1.6× Upsampling Rate	5.563×10^{-5}	1.058×10^{-4}	8.407×10^{-2}	0.975	0.960	0.936

break down at lower resolutions as evident in the 3.125 mm input configuration. We have also noticed a similar behaviour from the 3.125 mm input ablation model when it was provided with the 1.875 mm input. Compared to both models our multiple resolution approach demonstrates robustness in both configurations compared to a single resolution training approach.

Our out-of-distribution results for a higher upsampling rate have also demonstrated that the addition of a T1w input does improve the model output. We have observed a difference between our multimodal approach and unimodal approach that was statistically significant, for which we observed that every metric demonstrated a preference for the T1w model. However, the results also demonstrated a preference for the single resolution 3.125 mm input model compared to the multiple resolution model.

The results that we have obtained demonstrate differences in quality across our main model compared to our ablation models. In most of our experiments, we have observed an increase in upsampling quality on models that incorporated the T1w images compared to the unimodal ablation model. The ablation models that were evaluated in their native resolution usually performed the best at specified resolutions. Our multi-resolution training configuration with the inclusion of a high-resolution MRI modality achieved comparable results to the ablation models on higher quality DTI inputs and demonstrated better performance when the quality of the DTI inputs was lower ($> 2\,mm$ voxel size).

We have observed that the addition of T1w input does help in improving the quality of images in attempting to obtain an out of distribution target resolution. We observed a significantly better performance for our model that used a single upsampling rate of 1.6×. This indicates that the incorporation of a T1w image has been beneficial in the output quality at a previously unseen upsampling rate.

Figure 2 shows qualitative differences between the different model outputs on out-of-distribution images. We noticed certain features that resembled structures contained in the T1w image and not evident in the input dMRIs. In particular striations at fibre crossings between the Corpus Callosum and Corticospinal Tracts were sharper in the output of the multimodal model than in the ablation model.

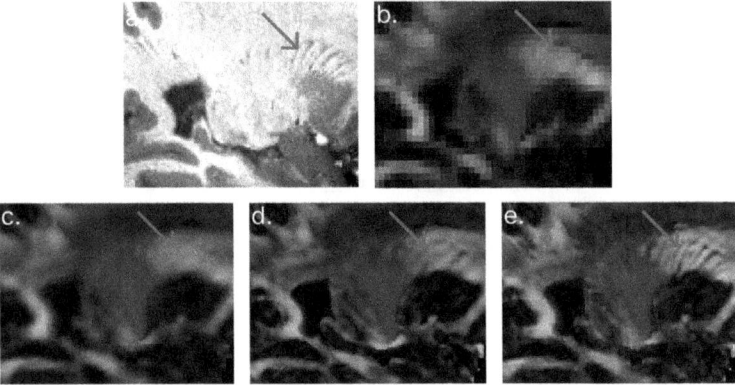

Fig. 2. Qualitative comparison of T1w input with colour FA maps constructed from competing upsampling methods. We show corresponding sagittal slices from (a) a 7T 0.7 mm T1w companion volume, (b) the original 2 mm DTI from the Connectom scanner, (c) the linearly upsampled DTI image to 0.7 mm, (d) a unimodal ablation model output, and (e) our multimodal network output. We highlight striations in internal capsule which are visible in the T1w image and gain higher definition from our multimodal approach.

5 Discussion

We have implemented an IQT model to upsample DTI data guided by high-resolution T1w images. Our tests showed that including the T1w input improved the output quality, especially in configurations where the DTI input had lower resolution and hence was less reliable. In addition, from our qualitative results we were able to view certain structures in our DTI output that were originally present in the T1w input and not evident in the DTI input. However, we have also noticed that while the addition of the T1w image improved the upsampling quality at every test we conducted, we have observed diminishing returns at higher DTI input resolutions. This further indicates that our model essentially uses structural information from the T1w input to incorporate into the DTI data when there is a lack of said structural information in the DTI input. Among the models including the T1w input, the ones trained on a single resolution closest to the test image performed the best in each experiment as it could be expected. The model trained with multiple resolutions, however, had a similar performance to the best one in each case and thus showed more robustness to different test cases.

To our knowledge, our approach is the first dMRI super resolution model that aims to combine multiple MRI modalities to reliably go beyond the resolution of its training set. Our approach, once qualitatively and quantitatively validated with the use of an ultra-high-resolution DTI scan, could be used in many applications to provide a reliable and quicker way to obtain high-resolution DTI scans. For example, transferring structural information of T1w scans from one scanner into diffusion scans acquired by a different scanner, as we demonstrated through our qualitative analysis. In addition to possibly improving the upsampling quality of dMRI scans, we speculate that the addition of a second MRI modality can be further used to reveal certain features that otherwise would not have been noticeable. For example, according to Wang et al. [23], certain brain lesions related to pharmacoresistant focal epilepsy such as Focal Cortical Dysplasia would be better detected in a 7T scanner because they are more clearly visible and have better-defined boundaries. By using high-resolution structural MRI images to upsample dMRI data, we our model could thus enhance the utility of dMRI for Focal Cortical Dysplasia. Similarly, the enhanced anatomical detail of the dMRI output of our model could find diagnostic or research applications in other medical conditions.

For future work, we plan to develop our fusion model by modifying the network architecture and training configuration to more effectively fuse features from the T1w input. Currently, each model is trained to a single output resolution, presenting the possibility that the model is more focused on features at this resolution than the T1w input. We are planning on mitigating this effect using multiple target resolutions, similar to the use of multiple resolutions on input. We are also planning on applying further data augmentations in the Diffusion Weighted Imaging (DWI) sequence to ensure the robustness of our model. Additionally, we are considering using different architectures that may fuse our data better such as by using attention layers [22], or with the use of a denoising diffusion model [7].

We plan to investigate the statistical distribution of our data and differences with ground truth to confirm that our model performs better than unimodal approaches while not introducing any bias to quantitative DTI metrics. Preliminary experiments on our qualitative test subject show a small decrease in MD and an increase in FA for our model prediction compared to the 2 mm input. On the HCP data, comparisons to ground truth found a small decrease in FA for subcortical and cerebral grey matter regions and an even smaller decrease in white matter regions; MD showed a slight increase in subcortical grey matter and white matter regions but no statistical significance for cerebral grey matter. These results could be explained as reduction of partial volumes effects going towards higher resolutions, but need to be investigated further.

Finally, our model is currently implemented to work with T1w and DTI data. We will investigate the use of other structural modalities, such as T2-weighted or FLAIR images, that could provide different and potentially useful information. We are also planning to extend the method to multi-shell dMRI techniques, e.g. Diffusion Kurtosis Imaging [10] or Mean Apparent Propagator MRI [17], that would allow a more advanced and specific charcterisation of brain microstructure.

To conclude, we have proposed an IQT model to combine information from a secondary MRI modality for upsampling dMRI models. Our proposed approach does demonstrate an increase in upsampling performance with the incorporation of a secondary MRI modality. Our plan is to develop this tool further by ensuring the dependence of our model on our T1w data with the use of different network architectures, or by eliminating possible ambiguity in our training configuration.

Acknowledgments. This study was funded by Wellcome Trust award 221915/Z/20/Z, MRC award MR/W031566/1 and was supported by the NIHR UCLH Biomedical Research Centre and the Medical Research Council (grant MR/W031566/1). Data were provided in part by the Human Connectome Project WU-Minn Consortium (Principal Investigators: David Van Essen and Kamil Ugurbil; 1U54MH091657), funded by the 16 NIH Institutes and Centers that support the NIH Blueprint for Neuroscience Research; and by the UK National Facility for In Vivo MR Imaging of Human Tissue Microstructure funded by the EPSRC (grant EP/M029778/1), and The Wolfson Foundation, and supported by a Wellcome Trust Strategic Award (104943/Z/14/Z).

Disclosure of Interests. The authors have no competing interests to declare that are relevant to the content of this article.

References

1. Alexander, D.C., et al.: Image quality transfer and applications in diffusion MRI. Neuroimage **152**, 283–298 (2017)
2. Basser, P.J., Mattiello, J., LeBihan, D.: MR diffusion tensor spectroscopy and imaging. Biophys. J . **66**(1), 259–267 (1994)

3. Billot, B., Greve, D.N., Van Leemput, K., Fischl, B., Iglesias, J.E., Dalca, A.V.: A learning strategy for contrast-agnostic MRI segmentation. In: Proceedings of Machine Learning Research, vol. 121, pp. 75–93 (2020)
4. Billot, B., Robinson, E., Dalca, A.V., Iglesias, J.E.: Partial volume segmentation of brain MRI scans of any resolution and contrast. In: Martel, A.L., et al. (eds.) MICCAI 2020. LNCS, vol. 12267, pp. 177–187. Springer, Cham (2020). https://doi.org/10.1007/978-3-030-59728-3_18
5. Boulahia, S.Y., Amamra, A., Madi, M.R., Daikh, S.: Early, intermediate and late fusion strategies for robust deep learning-based multimodal action recognition. Mach. Vis. Appl. **32**(6), 1–18 (2021). https://doi.org/10.1007/s00138-021-01249-8
6. Çiçek, Ö., Abdulkadir, A., Lienkamp, S.S., Brox, T., Ronneberger, O.: 3D U-Net: learning dense volumetric segmentation from sparse annotation. In: Ourselin, S., Joskowicz, L., Sabuncu, M.R., Unal, G., Wells, W. (eds.) MICCAI 2016. LNCS, vol. 9901, pp. 424–432. Springer, Cham (2016). https://doi.org/10.1007/978-3-319-46723-8_49
7. Ho, J., Jain, A., Abbeel, P.: Denoising diffusion probabilistic models. Adv. Neural. Inf. Process. Syst. **33**, 6840–6851 (2020)
8. Iglesias, J.E.: A ready-to-use machine learning tool for symmetric multi-modality registration of brain MRI. Sci. Rep. **13**(1), 6657 (2023)
9. Iglesias, J.E., Billot, B., Balbastre, Y., Tabari, A., Conklin, J., et al.: Neuroimage Joint super-resolution and synthesis of 1 mm isotropic MP-rage volumes from clinical MRI exams with scans of different orientation, resolution and contrast **237**, 118206 (2021)
10. Jensen, J.H., Helpern, J.A., Ramani, A., Lu, H., Kaczynski, K.: Diffusional kurtosis imaging: the quantification of non-gaussian water diffusion by means of magnetic resonance imaging. Magn. Resonance Med. Official J. Int. Soc. Magn. Resonance Med. **53**(6), 1432–1440 (2005)
11. Jones, D.K., Knösche, T.R., Turner, R.: White matter integrity, fiber count, and other fallacies: the do's and don'ts of diffusion MRI. Neuroimage **73**, 239–254 (2013)
12. Kim, S., Tregidgo, H.F.J., Eldaly, A.K., Figini, M., Alexander, D.C.: A 3D conditional diffusion model for image quality transfer – an application to low-field MRI (2023)
13. Kingma, D.P., Ba, J.: Adam: a method for stochastic optimization (2017)
14. Lin, H., Figini, M., D'Arco, F., Ogbole, G., Tanno, R., et al.: Low-field magnetic resonance image enhancement via stochastic image quality transfer. Med. Image Anal. **87**, 102807 (2023)
15. Mao, Y., Jiang, L., Chen, X., Li, C.: DisC-Diff: disentangled conditional diffusion model for multi-contrast MRI super-resolution. In: International Conference on Medical Image Computing and Computer-Assisted Intervention,. pp. 387–397. Springer (2023)
16. Ngiam, J., Khosla, A., Kim, M., Nam, J., Lee, H., Ng, A.Y.: Multimodal deep learning. In: Proceedings of the 28th International Conference on Machine Learning (ICML-11), pp. 689–696 (2011)
17. Özarslan, E., Koay, C.G., Shepherd, T.M., Komlosh, M.E., İrfanoğlu, M.O., et al.: Mean apparent propagator (MAP) MRI: a novel diffusion imaging method for mapping tissue microstructure. Neuroimage **78**, 16–32 (2013)
18. Pajevic, S., Pierpaoli, C.: Color schemes to represent the orientation of anisotropic tissues from diffusion tensor data: application to white matter fiber tract mapping in the human brain. Magn. Reson. Med. **43**(6), 921–921 (2000)

19. Steyaert, S., Pizurica, M., Nagaraj, D., Khandelwal, P., Hernandez-Boussard, T., et al.: Multimodal data fusion for cancer biomarker discovery with deep learning. Nat. Mach. Intell. **5**(4), 351–362 (2023)
20. Tanno, R., Worrall, D.E., Kaden, E., Ghosh, A., Grussu, F., et al.: Uncertainty modelling in deep learning for safer neuroimage enhancement: Demonstration in diffusion MRI. Neuroimage **225**, 117366 (2021)
21. Van Essen, D., Ugurbil, K., Auerbach, E., Barch, D., Behrens, T., et al.: The human connectome project: a data acquisition perspective. Neuroimage **62**(4), 2222–2231 (2012)
22. Vaswani, A., Shazeer, N., Parmar, N., Uszkoreit, J., Jones, L., et al.: Attention is all you need. In: Advances in Neural Information Processing Systems, vol. 30 (2017)
23. Wang, Z.I., Oh, S.H., Lowe, M., Larvie, M., Ruggieri, P., et al.: Radiological and clinical value of 7T MRI for evaluating 3T-visible lesions in pharmacoresistant focal epilepsies. Front. Neurol. **12** (2021)
24. Wang, Z., Bovik, A.: A universal image quality index. IEEE Signal Process. Lett. **9**(3), 81–84 (2002)

QID²: An Image-Conditioned Diffusion Model for *Q*-Space Up-Sampling of DWI Data

Zijian Chen[✉], Jueqi Wang, and Archana Venkataraman

Department of Electrical and Computer Engineering, Boston University, Boston, USA
{zijianc,jueqiw,archanav}@bu.edu

Abstract. We propose an image-conditioned diffusion model to estimate high angular resolution diffusion weighted imaging (DWI) from a low angular resolution acquisition. Our model, which we call QID², takes as input a set of low angular resolution DWI data and uses this information to estimate the DWI data associated with a target gradient direction. We leverage a U-Net architecture with cross-attention to preserve the positional information of the reference images, further guiding the target image generation. We train and evaluate QID² on single-shell DWI samples curated from the Human Connectome Project (HCP) dataset. Specifically, we sub-sample the HCP gradient directions to produce low angular resolution DWI data and train QID² to reconstruct the missing high angular resolution samples. We compare QID² with two state-of-the-art GAN models. Our results demonstrate that QID² not only achieves higher-quality generated images, but it consistently outperforms state-of-the-art baseline methods in downstream tensor estimation across multiple metrics and in generalizing to downsampling scenario during testing. Taken together, this study highlights the potential of diffusion models, and QID² in particular, for q-space up-sampling, thus offering a promising toolkit for clinical and research applications.

Keywords: Diffusion Weighted Imaging · Diffusion Models · Deep Learning · Q-Space Up-sampling · Tensor Reconstruction

1 Introduction

Diffusion weighted imaging (DWI) is a non-invasive technique that capitalizes on the directional diffusivity of water to probe the tissue microstructure of the brain [4]. A typical DWI acquisition applies multiple magnetic gradients, with the field strength controlled by the b-value and the gradient directions given by the b-vectors. Mathematically, these gradients can be represented by a set of coordinates on the sphere, where the magnitude and direction of each coordinate is related to the corresponding b-value and b-vector, respectively. The

---

Z. Chen and J. Wang—Equal Contribution.

domain of all such coordinates is called the q-space [28]. In general, a denser sampling of directions in the q-space, also known as the angular resolution, leads to higher quality DWI. For example, higher angular resolution acquisitions can improve the tensor estimation [12] and facilitates the progression from single-tensor models [1] to constrained spherical deconvolution models [23] that estimate a fiber orientation distribution function (fODF), which captures more complex fiber configurations. However, increasing the angular resolution also prolongs the acquisition time, which can be impractical in clinical settings. Not only are longer acquisitions more expensive, but they are also difficult for some patients to tolerate, which in turn increases the risk of artifacts due to subject motion [10]. Given these challenges, it is necessary to explore computational methods that can achieve high-quality DWI with a minimal number of initial scan directions.

Several studies have applied generative deep learning to DWI data. For example, the work of [29] uses a spherical U-Net to directly estimate the ODF using DWI acquired with only 60 gradient directions. More recently, generative adversarial networks (GANs) have also been used to estimate DWI volumes. Specifically, the work of [14] generates DWI for a user-specified gradient direction based on a combination of T1 and T2 images. Similarly, the authors of [22] use the Pix2Pix model introduced by [7] to synthesize DWI with 6 gradient directions from data originally captured with only 3 gradient directions [22]. Further variants of the GAN model, such as CycleGAN and DC^2Anet, have been applied to simulate a high b-value image from a low b-value one [15]. Beyond GANs, autoencoders have also been used to adjust the apparent b-value [8]. While these works are seminal contributions to the field, none of them consider the clinically relevant problem of up-sampling a low angular resolution DWI acquisition.

Diffusion models have emerged as powerful tool for image generation. At a high level, they work by successively adding Gaussian noise to the input and then learning to reverse this noising process [6]. Diffusion models have been employed in several medical imaging tasks, including image translation between modalities [11], super-resolution and artifact removal [26], registration [9], and segmentation [13]. We will leverage diffusion models to up-sample the DWI gradient directions, which to our knowledge, has not been explored in prior work.

In this paper, we propose an image-conditioned diffusion model, which we call QID^2, that can estimate high angular resolution DWI data from a low angular resolution acquisition[1]. One highlight is that QID^2 automatically identifies several closest available gradient directions and uses the corresponding images as prior knowledge for generating images from any target direction not included in the initial scan. This target image generation process, carried out using a U-Net based structure conditioned on this prior information, can be seen as an extrapolation based on the identified directions and images. By focusing on the most relevant data, QID^2 solicits more targeted prior information and is more computationally efficient. We train and evaluate QID^2 on DWI curated from the

---

[1] Source code for our model is available online at https://github.com/jueqiw/ Diffusion-Model-for-Up-sampling-Diffusion-Weighted-Imaging.

Human Connectome Project (HCP) dataset [24]. Our model demonstrates superior performance over GAN-based approaches, particularly when the available low angular resolution images are sparsely distributed across the sphere.

2 Methods

Figure 1 provides an overview of our QID^2 framework. For any user-specified target gradient $\mathbf{b}_g$, our model will find and take as input the R closest reference b-vectors $\bar{\mathbf{b}} = (\mathbf{b}_1, \ldots, \mathbf{b}_R)$ available in the low angular resolution scan and the corresponding DWI slices $\bar{\mathbf{X}} = (\mathbf{X}_1, \ldots, \mathbf{X}_R)$. QID^2 will then output the estimated target image $\mathbf{X}_{\mathbf{b}_g}$. We can obtain a high angular resolution DWI by sweeping the target gradient directions across the sphere and aggregating the generated images with the original low angular resolution scan.

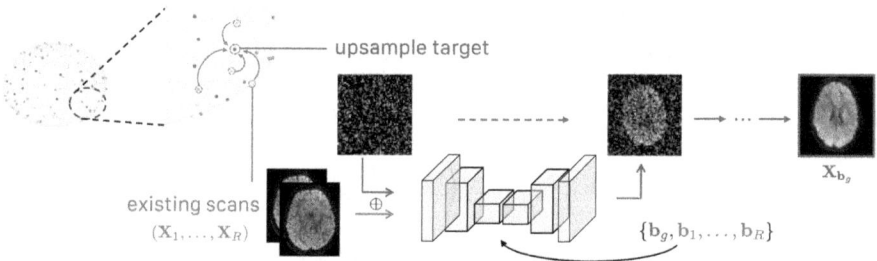

Fig. 1. QID^2 framework for up-sampling the angular resolution of DWI. The gray sphere represents the q-space. Red marks are the directions in the low angular resolution scan, and blue marks are the target gradient directions for image generation. (Color figure online)

2.1 A Diffusion Model for Q-Space Up-Sampling of DWI

Inspired by recently-introduced image-conditioned Denoising Diffusion Probabilistic Models (DDPMs) [25], we design a position-aware diffusion model that leverages "neighboring" DWI data to estimate the image associated with a target gradient direction. Similar to traditional diffusion models [6], QID^2 is comprised of both a forward noising process and a reverse denoising process.

In the forward process, Gaussian noises are added successively at each time step $t \in \{0, 1, \ldots, T\}$ to the generated image. This corruption process is

$$q\left(\mathbf{X}_{\mathbf{b}_g}^{(t)} \mid \mathbf{X}_{\mathbf{b}_g}^{(t-1)}\right) = \mathcal{N}\left(\mathbf{X}_{\mathbf{b}_g}^{(t)}; \sqrt{1-\beta_t}\mathbf{X}_{\mathbf{b}_g}^{(t-1)}, \beta_t \mathbf{I}\right), \quad t \geq 1, \qquad (1)$$

where $\{\beta_t\}$ are the forward process variances, and $\mathbf{X}_{\mathbf{b}_g}^{(t)}$ is the noisy image at time t. By repeatedly applying Eq. (1) to the starting image $\mathbf{X}_{\mathbf{b}_g}^{(0)}$, we have

$$q\left(\mathbf{X}_{\mathbf{b}_g}^{(t)} \mid \mathbf{X}_{\mathbf{b}_g}^{(0)}\right) = \mathcal{N}\left(\mathbf{X}_{\mathbf{b}_g}^{(t)}\ \sqrt{\overline{\alpha}_t}\mathbf{X}_{\mathbf{b}_g}^{(0)},\ (1-\overline{\alpha}_t)\mathbf{I}\right) \qquad (2)$$

where $\alpha_t = 1 - \beta_t$ and $\overline{\alpha}_t = \prod_{s=1}^{t} \alpha_s$. Therefore, at step t, the generated image $\mathbf{X}_{\mathbf{b}_g}^{(t)}$ can be represented as a function of the initialization $\mathbf{X}_{\mathbf{b}_g}^{(0)}$:

$$\mathbf{X}_{\mathbf{b}_g}^{(t)} = \sqrt{\overline{\alpha}_t}\,\mathbf{X}_{\mathbf{b}_g}^{(0)} + \sqrt{1-\overline{\alpha}_t}\,\epsilon, \qquad \epsilon \sim \mathcal{N}(0,\mathbf{I}). \qquad (3)$$

While the forward noising process operates solely on $\mathbf{X}_{\mathbf{b}_g}^{(0)}$, the reference DWI slices $\{\mathbf{X}_1, \ldots, \mathbf{X}_R\}$ will be used to guide the subsequent denoising process. Rather than constructing a separate network to encode the reference images, which greatly increases the number of parameters and may introduce information loss, we opt to simply concatenate these slices with the target image being generated (i.e., denoised) at each time t as $\bar{\mathbf{X}}_{\mathbf{b}_g}^{(t)} = \mathbf{X}_{\mathbf{b}_g}^{(t)} \bigoplus_{i=1}^{R} \mathbf{X}_{\mathbf{b}_i}$.

Starting from the fully corrupted image $\bar{\mathbf{X}}_{\mathbf{b}_g}^{(T)}$, the reverse process aims to gradually recover the original image $\bar{\mathbf{X}}_{\mathbf{b}_g}^{(0)}$. We denote this process as $p_\theta(\cdot)$, where θ denotes the learnable parameters of the underlying neural network. By restricting the denoising to be Gaussian, the process $p_\theta(\cdot)$ can be written:

$$p_\theta\left(\bar{\mathbf{X}}_{\mathbf{b}_g}^{(t-1)} \mid \bar{\mathbf{X}}_{\mathbf{b}_g}^{(t)}; \{\mathbf{b}_g, \bar{\mathbf{b}}\}\right) = \mathcal{N}\left(\bar{\mathbf{X}}_{\mathbf{b}_g}^{(t-1)}; \mu_\theta\left(\bar{\mathbf{X}}_{\mathbf{b}_g}^{(t)}, \{\mathbf{b}_g, \bar{\mathbf{b}}\}\right), \sigma_t^2 \mathbf{I}\right), \qquad (4)$$

where the variances σ_t^2 are hyperparameters of the model. We note that the denoising process relies on the references DWI data $\{\mathbf{X}_1, \ldots, \mathbf{X}_R\}$ and the corresponding gradient directions $\bar{\mathbf{b}} = \{\mathbf{b}_1, \ldots, \mathbf{b}_R\}$, and the target direction $\mathbf{b}_g$. This combination of inputs allows QID^2 to be position-aware.

To reverse the forward noising process, we train QID^2 by minimizing the KL-divergence between $p_\theta(\cdot)$ and $q(\cdot)$ at each time step t. As shown in [6] this loss minimization is equivalent to matching the mean functions, i.e.,

$$\mathcal{L} = \mathbb{E}_{t,q}\left[\left\|\frac{1}{\sqrt{\alpha_t}}\left(\bar{\mathbf{X}}_{\mathbf{b}_g}^{(t)} - \frac{\beta_t}{\sqrt{1-\bar{\alpha}_t}}\epsilon\right) - \mu_\theta\left(\bar{\mathbf{X}}_{\mathbf{b}_g}^{(t)}, \{\mathbf{b}_g, \bar{\mathbf{b}}\}\right)\right\|^2\right]. \qquad (5)$$

The mean function $\mu_\theta(\cdot)$ is generated with a U-Net architecture [17] with the cross-attention mechanism based on the concatenated gradient vectors $\mathbf{b} = [\mathbf{b}_g\ \mathbf{b}_1\ \cdots\ \mathbf{b}_R]$. Specifically, the encoding block is computed as follows:

$$\mathbf{H}_1 = \mathrm{FF}(\mathbf{H}_0) + \mathbf{H}_0, \quad \mathbf{H}_2 = \mathrm{Attn}(\mathbf{H}_1, \mathbf{b}) + \mathbf{H}_1,$$

where $\mathrm{FF}(\cdot)$ denotes a feed-forward network, $\mathbf{H}_0$ denotes the block input, and

$$\mathrm{Attn}(\mathbf{H}_1, \mathbf{b}) = \mathrm{Softmax}\left(\frac{(W_Q \mathbf{H}_1)(W_K \mathbf{b})^\top}{\sqrt{d_k}}\right) W_V \mathbf{b},$$

with W_Q, W_K, W_V being the learned weights and d_k being the dimension of $\mathbf{b}$. The decoding block follows a similar expression but includes skip connections from the corresponding encoding block. This design ensures that image features are effectively attended to and integrated with positional information.

Once QID^2 is trained, we can generate DWI for arbitrary gradient directions by sampling from the standard normal distribution and applying the reverse process in Eq. (4) recursively with the corresponding reference images, namely:

$$\bar{\mathbf{X}}_{\mathbf{b}_g}^{(t-1)} = \mu_\theta \left(\bar{\mathbf{X}}_{\mathbf{b}_g}^{(t)}, \{\mathbf{b}_g, \bar{\mathbf{b}}\} \right) + \sigma_t \epsilon, \qquad \epsilon \sim \mathcal{N}(0, \mathbf{I}). \tag{6}$$

2.2 Baseline Comparison Methods

We compare QID^2 with two state-of-the-art GAN models. The first model is a conditional GAN (**cGAN**) for image generation proposed by [7]. We use the same cross-attention U-Net architecture for the generator as used in QID^2. We use a PatchGAN discriminator [7] and inject the gradient direction information $\{\mathbf{b}_g, \bar{\mathbf{b}}\}$ into the discriminator with cross-attention mechanism. We train the generator to minimize the GAN objective plus a regularization term that encourages voxel-level similarity of the generated and ground-truth images:

$$G^* = \arg \min_G \max_D \lambda_G \left[\mathbb{E}_{\mathbf{X}} \left[\log D(\mathbf{X}, \mathbf{b}) \right] + \mathbb{E}_{\widetilde{\mathbf{X}}} \left[\log(1 - D(\widetilde{\mathbf{X}}, \mathbf{b})) \right] \right] + \lambda_V \mathcal{L}_1(G)$$

where $\mathbf{X} = [\mathbf{X}_{\mathbf{b}_g}, \mathbf{X}_1, \ldots, \mathbf{X}_R]$ is the concatenated real sample with $\mathbf{X}_{\mathbf{b}_g}$ drawn from the (high resolution) training data and $\widetilde{\mathbf{X}} = [G(\mathbf{X}_{1:R}, \mathbf{b}), \mathbf{X}_1, \ldots, \mathbf{X}_R]$ represents the synthesized data of generated DWI and real reference slices. Finally, λ_G and λ_V balance the adversarial and similarity L_1 losses, respectively.

The second model is the Q-space conditional GAN (**qGAN**) proposed by [14]. Unlike QID^2 and the cGAN baseline, qGAN incorporates the gradient directions and reference DWI data using a feature-wise linear modulation scheme. The qGAN discriminator is also a conditional U-Net and combines the gradient directions and reference DWI data via an inner product. Although the inputs to the original qGAN model [14] are a single structural image (e.g., B0, T1, T2) and a user-defined target gradient, we provide the same set of closest directions and corresponding images as input to ensure a fair comparison with QID^2.

Finally, as a sanity check, we compare the deep learning models to a simple interpolation scheme (**Interp**), in which we express the target gradient direction as a linear combination of the reference gradients and then use the linear coefficients to interpolate between the reference DWI slices to obtain the target.

2.3 Implementation Details

For QID^2, we use a linear noise schedule of 1000 time steps. The QID^2 U-Net employs [128, 128, 256] channels across three levels with one residual block per level. We use the Adam optimizer with a learning rate of 2.5×10^{-5}, $\beta_1 = 0.5$ and $\beta_2 = 0.999$. These hyperparameters are selected based on a relevant study [16] and not fine-tuned. We use the same U-Net architecture for the cGAN generator with the same set of hyperparameters. We fix $\lambda_G = 1$ and $\lambda_V = 100$ in the loss function weights for both GAN methods. The discriminator is updated once for every two updates of the generator during training [14]. For both qGAN and cGAN, we use a learning rate of 5×10^{-5} with the Adam optimizer. We evaluate all models with both $R = 3$ and $R = 6$ reference DWI data. To avoid memory issues, we train the deep learning models to generate 2D axial slices, which we stack into 3D DWI volumes. Each 2D image has a size of $(145, 174)$. During training, we independently normalize each slice from its original intensity to a range of $[0, 1]$. Data augmentation is employed to enhance model training. Specifically, we use rotations by random angles in $[-15°, 15°]$ and random spatial scaling factors in $[0.9, 1.1]$. The final output is rescaled voxel-wise back to the original intensity and masked by the subject $\mathbf{X}_{\mathbf{b}_g}$ image.

3 Experimental Results

Dataset Curation and Preprocessing: We curate a total of 720 subjects from the HCP S1200 release [24]. The remaining HCP subjects are excluded due to an inconsistent number of gradient directions at $b = 1000$ s/mm². The DWI is acquired on a Siemens 3T Connectome scanner at 3 shells ($b = 1000, 2000$ and $3000 \ s/mm^2$). Each shell has exactly 90 gradient directions sampled uniformly on the sphere. The voxel size is $1.25 \times 1.25 \times 1.25$ mm³. The data is preprocessed with distortion/motion removal and registration to the 1.25 mm structural space.

Clinical diffusion imaging typically uses lower b-values with approximately 30 gradient directions [4]. To better accommodate this situation, we focus our evaluation on the $b = 1000$ s/mm² shell. From here, we construct low angular resolution DWI data by subsampling the 90 gradient directions to 30 evenly spaced ones that preserve the uniformity of the sphere [2]. The data for the remaining 60 directions serve as the targets for model training and evaluation.

Each volume is broken down into 145 axial slices. The deep learning models are trained to predict the image slices for each target direction. Based on this scheme, we create 60 samples for each slice. Each sample consists of one 2D slice for the target gradient direction and R reference slices corresponding to the closest low resolution gradient directions. The distance between gradients is defined by the geodesic distance on the sphere: $d(\mathbf{b}_1, \mathbf{b}_2) = \arccos(\mathbf{b}_1 \mathbf{b}_2^\top)$ [3].

Finally, we use 576 HCP subjects for training, 72 for validation, and 30 for testing. The original DWI scans are treated as the gold standard for evaluation.

Comparing Reconstructed Image Quality: Figure 2 presents qualitative results that compare the ground-truth DWIs to those generated by QID² and the baseline methods. The GAN models and Interp fail to preserve high-frequency details in the synthesized DWI data, while QID² succeeds in capturing the finer details more accurately, as highlighted in the zoomed-in blue boxes.

Table 1 (left) reports the Fréchet inception distance (FID) [5] and the structural similarity index measure (SSIM) [27] of the synthesized DWI data. Specifically, FID measures the realism and diversity of images by comparing the feature distributions between the generated and ground truth ones, while SSIM quantifies the similarity based on luminance, contrast, and structural information.

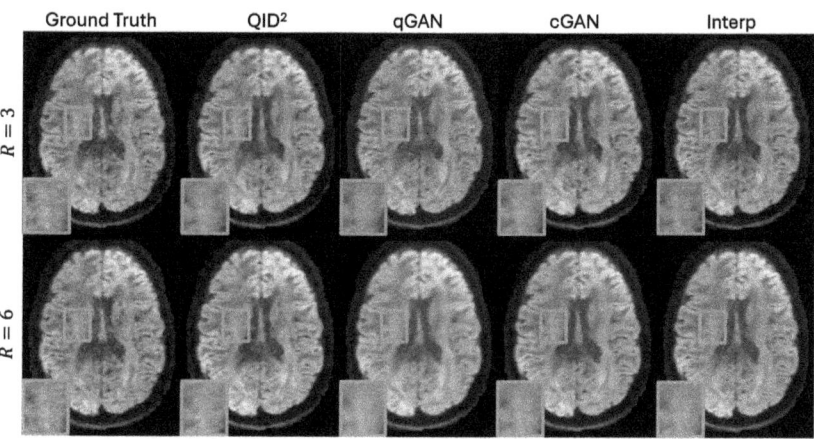

Fig. 2. Qualitative results that compare the ground-truth DWI acquisition to images generated by QID^2 and the baselines methods for $R = 3$ and $R = 6$. Zoomed-in area highlights details that are preserved by our method and do not appear in the baselines.

We observe that QID^2 achieves nearly a two-fold improvement (i.e., decrease) in FID than the GAN models for both $R = 3$ and $R = 6$, which indicates that the DWI data generated by diffusion possess higher quality and greater diversity. Although the GAN models achieves a slightly higher SSIM than QID^2, the difference is not statistically significant using a two-sample (paired) t-test. Interestingly, the simple interpolation technique achieves better FID than QID^2 when $R = 3$. This is likely because the interpolation tracks the closest reference image, which is more akin to the original DWI distribution. However, the improved FID does not generalize to better tensor estimation, as seen in the next section.

Table 1. Quantitative evaluation of the generated image quality (left) and FA estimation quality (right) of QID^2, the GAN models, and interpolation for different R. The best performance of each metric is highlighted in bold.

Methods	Image FID ↓	Image SSIM ↑	FA Error ↓	FA Map SSIM ↑
QID^2 (R=3)	14.07	0.895 ± 0.045	$\mathbf{0.027 \pm 0.003}$	$\mathbf{0.866 \pm 0.043}$
qGAN(R=3)	24.85	0.893 ± 0.046	0.037 ± 0.002	0.792 ± 0.052
cGAN(R=3)	29.93	0.913 ± 0.039	0.099 ± 0.014	0.643 ± 0.159
Interp(R=3)	**8.96**	$\mathbf{0.917 \pm 0.038}$	0.057 ± 0.026	0.750 ± 0.110
QID^2 (R=6)	**16.29**	0.900 ± 0.045	$\mathbf{0.027 \pm 0.003}$	$\mathbf{0.863 \pm 0.042}$
qGAN(R=6)	71.44	0.905 ± 0.042	0.040 ± 0.004	0.801 ± 0.045
cGAN(R=6)	36.33	0.915 ± 0.037	0.031 ± 0.004	0.851 ± 0.044
Interp(R=6)	21.46	$\mathbf{0.933 \pm 0.044}$	0.038 ± 0.006	0.815 ± 0.052

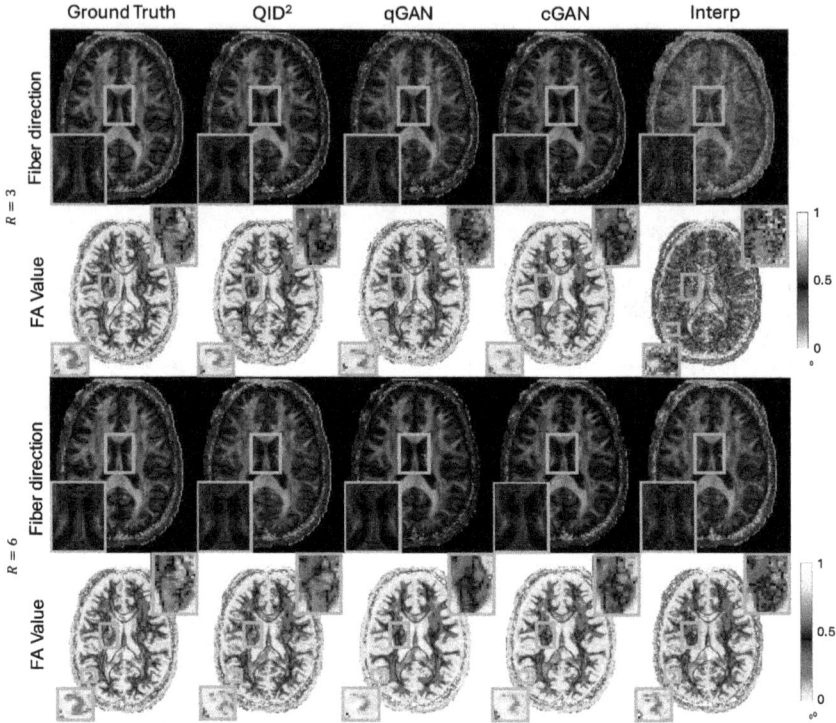

Fig. 3. Qualitative comparison between the ground truth and estimated images. **Row 1/3:** Colored fiber orientation maps with minimal visually detectable differences among the images. Zoomed-out regions show the estimated tensors in the orange box area. **Row 2/4:** FA value maps, where brighter colors indicate higher FA values. Significant differences compared to the ground truth are zoomed-out with orange boxes.

Impact on Tensor Estimation: We estimate the fractional anisotropy (FA) using the standard single-tensor model [1]. Figure 3 shows the fiber direction and FA value maps among the ground-truth, QID^2 and baseline methods for $R = 3$ and $R = 6$. Similarly to the finding in the reconstructed image, we observe that the qGAN and cGAN methods capture the general FA trends but fail to capture the high frequency features. Conversely, the diffusion-generated image by QID^2 more closely resembles the ground-truth data by capturing finer details more accurately. This shows that the visual differences in the reconstructed images in Fig. 2 are important when estimating tensors. The Interp method fails to generate realistic FA maps for $R = 3$. Empirically, we also observe quality issues with Interp for $R = 6$ even though they are less evident in the figure.

Table 1 (right) reports the mean absolute error and SSIM, as compared to the FA computed from the ground-truth high angular resolution DWI. As seen, QID^2 consistently outperforms the GAN-based model and the Interp method for both $R = 3$ and $R = 6$. Specifically for $R = 3$, the error in FA is roughly three times lower for QID^2 than for the GANs. QID^2 also achieves significantly higher SSIM values. These trends persist when the number of reference images increases to $R = 6$, i.e., even when more prior information is provided. However, the relative performance gain over the GANs shrink. Additionally, although the image-based metrics are better for the interpolation-generated (Interp) images, QID^2 outperforms this baseline by a large margin when estimating FA. Taken together, these results suggest that QID^2 is particularly effective in scenarios where the images are scarce and distributed sparsely, i.e., smaller values of R.

Fewer Available Initial Directions: As a final evaluation, we apply the models trained when assuming 30 initial gradient directions to testing data for which fewer gradient directions are available. Table 2 compares the performance of QID^2 and the baseline methods for 20 (top) and 10 (bottom) uniformly-distributed gradient directions at test time. Intuitively, we observe that the performance of all models worsens as the number of initial gradient directions decreases. However, QID^2 remains relatively stable from 20 to 10 directions, while baseline models experience a sharper drop. Similar to Table 1, QID^2 achieves comparable (but not the best) performance in reconstructed image quality, but it leads by a large margin with respect to tensor estimation. This result further highlights the benefits of QID^2, as even though QID^2 is trained on a dense gradient distribution, it can generalize effectively when applied to a sparser one.

Table 2. Quantitative evaluation of the generated image quality (left) and FA estimation quality (right) of each model when only 20 and 10 initial gradient directions are available at test time. The best performance of each metric is highlighted in bold.

	Methods	Image FID ↓	Image SSIM ↑	FA Error ↓	FA Map SSIM ↑
20 initial directions	$QID^2(R=3)$	15.28	0.883±0.048	**0.029±0.003**	**0.851±0.048**
	qGAN($R=3$)	24.75	0.886±0.04	0.056±0.003	0.624±0.067
	cGAN($R=3$)	29.75	0.904±0.042	0.102±0.014	0.622±0.153
	Interp($R=3$)	**9.91**	**0.934±0.042**	0.083±0.033	0.676±0.123
	$QID^2(R=6)$	16.91	0.887±0.049	**0.030±0.003**	**0.843±0.048**
	qGAN($R=6$)	71.94	0.896±0.045	0.100±0.025	0.580±0.086
	cGAN($R=6$)	36.31	0.906±0.041	0.073±0.011	0.591±0.096
	Interp($R=6$)	19.86	**0.911±0.041**	0.050±0.011	0.757±0.074
10 initial directions	$QID^2(R=3)$	14.68	0.882±0.054	**0.037±0.004**	**0.798±0.069**
	qGAN($R=3$)	23.74	0.881±0.054	0.060±0.004	0.591±0.078
	cGAN($R=3$)	30.27	0.899±0.048	0.111±0.019	0.552±0.143
	Interp($R=3$)	**10.65**	**0.914±0.051**	0.112±0.033	0.613±0.127
	$QID^2(R=6)$	17.48	0.883±0.056	**0.038±0.004**	**0.783±0.072**
	qGAN($R=6$)	72.55	0.890±0.052	0.120±0.030	0.521±0.113
	cGAN($R=6$)	36.45	**0.898±0.048**	0.083±0.014	0.507±0.120
	Interp($R=6$)	**15.73**	0.880±0.057	0.091±0.023	0.643±0.110

4 Discussion and Future Work

Our proposed image-conditioned diffusion model, named QID^2, is designed to upsample a low angular resolution DWI acquisition to have a higher angular resolution. One key innovation is our use of the automatically identified neighboring gradient directions in the low angular resolution DWI as prior information to improve the generation of images associated with new directions. We also propose an efficient way to encode this prior information (images and corresponding gradient directions) that ensures the effective attention and integration between the two. Our real-world experiments demonstrate that QID^2 outperforms two baseline GAN models in both image quality and tensor estimation.

However, our study is not without limitations. A notable one is that the models are trained and validated on a single dataset, namely HCP. While the HCP dataset provides a large number of subjects and a state-of-the-art acquisition protocol that can be used to create both low and high angular resolution DWI for model training, it may not fully capture the types of imaging acquisitions used in a clinical setting. To address this issue, future work should train and test QID^2 on additional datasets with different scanning protocols and patient characteristics (e.g., brain lesions). An example that can be used in future work is the CDMRI Quantitative Connectivity (QuantConn) challenge dataset [20].

A second limitation of our study is the focus on a single-shell reconstruction, and correspondingly, a single tensor estimation. While the use of $b = 1000$ s/mm^2 images in this work aligns with common clinical practices [4], it only enables us to fit a basic tensor model and is insufficient for more sophisticated fODF estimation and tractography analysis. Though it is not standard in many clinical workflows due to the difficulties of acquisition, tractography is still a useful tool to optimize surgical planning and postoperative assessment for tumors and vascular malformations [4]. Our current QID^2 can be adapted to multi-shell image upsampling, but it would require training a separate model for each shell. Future work will incorporate the b-value as an auxiliary input to QID^2 to enhance efficiency and streamline the upsampling process across multiple shells.

Despite these limitations, we believe that QID^2 has promise in clinical applications. Here, clinical deployment would follow a two-step procedure. First, we can pretrain QID^2 on a large collection of publicly available datasets, from HCP to the QuantConn challenge dataset to UK Biobank. Second, individual sites can opt to fine-tune this base model using their in-house datasets in order to match the specific MRI scanning protocols and patient conditions under evaluation. On the user (clinician) end, a typical workflow using the deployed model includes performing a standard low angular DWI scan, specifying target directions (not necessarily uniform), followed by processing the data with QID^2 to generate high-angular images, and combining the outputs for downstream analysis. This approach will help reduce scan times and mitigate patient discomfort, which are both significant challenges in the clinical use of DWI [21].

5 Conclusion

We introduce an image-conditioned diffusion model (QID^2) that can generate high angular resolution DWI from low angular resolution data, effectively estimating high-quality imaging with limited initial scan directions. Our approach takes advantage of similar DWI data as prior information to predict the data for any user-specified gradient direction. The results demonstrate that diffusion-generated DWIs by QID^2 achieve superior quality and significantly outperform those generated by baseline models in downstream tensor modeling tasks. QID^2 only shows a slight performance drop when applying to a sparser initial directions distribution during testing, demonstrating its superior generalizability. Although our method currently exhibits longer training times due to the denoising characteristics of DDPMs, this limitation could be mitigated by employing more efficient sampling techniques [19] and one-shot training [18] in future work.

Acknowledgements. This work was supported by the National Institutes of Health R01 HD108790 (PI Venkataraman), the National Institutes of Health R01 EB029977 (PI Caffo), the National Institutes of Health R21 CA263804 (PI Venkataraman).

References

1. Basser, P.J., Mattiello, J., LeBihan, D.: MR diffusion tensor spectroscopy and imaging. Biophys. J . **66**(1), 259–267 (1994)
2. Cheng, J., Shen, D., Yap, P.T., Basser, P.J.: Single-and multiple-shell uniform sampling schemes for diffusion MRI using spherical codes. IEEE Trans. Med. Imaging **37**(1), 185–199 (2017)
3. Chung, M.K., Chen, Z.: Embedding of functional human brain networks on a sphere. arXiv preprint arXiv:2204.03653 (2022)
4. Doshi, A., Gerke, L., Marchione, J., Bou-Haidar, P., Delman, B.: Physiologic evaluation of the brain with magnetic resonance imaging. Youmans and Winn Neurological Surgery. 7th ed. New York: Elsevier, pp. 69–95 (2017)
5. Heusel, M., Ramsauer, H., Unterthiner, T., Nessler, B., Hochreiter, S.: GANs trained by a two time-scale update rule converge to a local nash equilibrium. In: Advances in Neural Information Processing Systems, vol. **30** (2017)
6. Ho, J., Jain, A., Abbeel, P.: Denoising diffusion probabilistic models. Adv. Neural. Inf. Process. Syst. **33**, 6840–6851 (2020)
7. Isola, P., Zhu, J.Y., Zhou, T., Efros, A.A.: Image-to-image translation with conditional adversarial networks. In: Proceedings of the IEEE Conference on Computer Vision and Pattern Recognition, pp. 1125–1134 (2017)
8. Jha, R.R., Jaswal, G., Bhavsar, A., Nigam, A.: Single-shell to multi-shell dMRI transformation using spatial and volumetric multilevel hierarchical reconstruction framework. Magn. Reson. Imaging **87**, 133–156 (2022)
9. Kim, B., Han, I., Ye, J.C.: DiffuseMorph: unsupervised deformable image registration using diffusion model. In: European Conference on Computer Vision, pp. 347–364. Springer (2022)
10. Koh, D.M., Collins, D.J.: Diffusion-weighted MRI in the body: applications and challenges in oncology. Am. J. Roentgenol. **188**(6), 1622–1635 (2007)

11. Li, Y., et al.: Zero-shot medical image translation via frequency-guided diffusion models. IEEE Trans. Med. Imaging (2023)
12. Michailovich, O., Rathi, Y.: Fast and accurate reconstruction of HARDI data using compressed sensing. In: International Conference on Medical Image Computing and Computer-assisted Intervention, pp. 607–614. Springer (2010)
13. Rahman, A., Valanarasu, J.M.J., Hacihaliloglu, I., Patel, V.M.: Ambiguous medical image segmentation using diffusion models. In: Proceedings of the IEEE/CVF Conference on Computer Vision and Pattern Recognition, pp. 11536–11546 (2023)
14. Ren, M., Kim, H., Dey, N., Gerig, G.: Q-space conditioned translation networks for directional synthesis of diffusion weighted images from multi-modal structural MRI. In: de Bruijne, M., et al. (eds.) MICCAI 2021. LNCS, vol. 12907, pp. 530–540. Springer, Cham (2021). https://doi.org/10.1007/978-3-030-87234-2_50
15. Rezaeijo, S.M., Entezari Zarch, H., Mojtahedi, H., Chegeni, N., Danyaei, A.: Feasibility study of synthetic DW-MR images with different b values compared with real DW-MR images: quantitative assessment of three models based-deep learning including CycleGAN, Pix2PiX, and DC2Anet. Appl. Magn. Reson. **53**(10), 1407–1429 (2022)
16. Rombach, R., Blattmann, A., Lorenz, D., Esser, P., Ommer, B.: High-resolution image synthesis with latent diffusion models. In: Proceedings of the IEEE/CVF Conference on Computer Vision and Pattern Recognition, pp. 10684–10695 (2022)
17. Ronneberger, O., Fischer, P., Brox, T.: U-Net: convolutional networks for biomedical image segmentation. In: Medical Image Computing and Computer-Assisted Intervention–MICCAI 2015: 18th International Conference, Munich, Germany, October 5-9, 2015, Proceedings, Part III 18, pp. 234–241. Springer (2015)
18. Shaham, T.R., Dekel, T., Michaeli, T.: SinGAN: learning a generative model from a single natural image. In: Proceedings of the IEEE/CVF International Conference on Computer Vision, pp. 4570–4580 (2019)
19. Song, J., Meng, C., Ermon, S.: Denoising diffusion implicit models. arXiv preprint arXiv:2010.02502 (2020)
20. Strike, L.T., et al.: Queensland Twin Imaging (QTIM) (2023). https://doi.org/10.18112/openneuro.ds004169.v1.0.7
21. Tae, W.S., Ham, B.J., Pyun, S.B., Kang, S.H., Kim, B.J.: Current clinical applications of diffusion-tensor imaging in neurological disorders. J. Clin. Neurol. **14**(2), 129–140 (2018)
22. Tatekawa, H., et al.: Deep learning-based diffusion tensor image generation model: a proof-of-concept study. Sci. Rep. **14**(1), 2911 (2024)
23. Tournier, J.D., Calamante, F., Connelly, A.: Robust determination of the fibre orientation distribution in diffusion MRI: non-negativity constrained super-resolved spherical deconvolution. Neuroimage **35**(4), 1459–1472 (2007)
24. Van Essen, D.C., et al.: The WU-Minn human connectome project: an overview. Neuroimage **80**, 62–79 (2013)
25. Waibel, D.J., Röell, E., Rieck, B., Giryes, R., Marr, C.: A diffusion model predicts 3D shapes from 2D microscopy images. In: 2023 IEEE 20th International Symposium on Biomedical Imaging (ISBI), pp. 1–5. IEEE (2023)
26. Wang, J., Levman, J., Pinaya, W.H.L., Tudosiu, P.D., Cardoso, M.J., Marinescu, R.: InverseSR: 3D brain MRI super-resolution using a latent diffusion model. In: International Conference on Medical Image Computing and Computer-Assisted Intervention, pp. 438–447. Springer (2023)
27. Wang, Z., Bovik, A., Sheikh, H., Simoncelli, E.: Image quality assessment: from error visibility to structural similarity. IEEE Trans. Image Process. **13**(4), 600–612 (2004)

28. Yeh, F.C., Irimia, A., de Almeida Bastos, D.C., Golby, A.J.: Tractography methods and findings in brain tumors and traumatic brain injury. Neuroimage **245**, 118651 (2021)
29. Zhao, H., Deng, C., Wang, Y., Ma, J.: Better fibre orientation estimation with single-shell diffusion MRI using spherical U-Net. In: International Conference of Pioneering Computer Scientists, Engineers and Educators, pp. 3–12. Springer (2023)

Ts-FWE: Token-Aware Single-Shell Free Water Estimation for Brain Diffusion MRI

Tianyuan Yao[1], Derek Archer[2], Zhiyuan Li[1], Leon Y. Cai[1], Praitayini Kanakaraj[1], Nancy Newlin[1], Quan Liu[1], Ruining Deng[1], Can Cui[1], Shunxing Bao[1], Kurt Schilling[2], Bennett A. Landman[1], and Yuankai Huo[1](✉)

[1] Vanderbilt University, Nashville, TN 37215, USA
yuankai.huo@vanderbilt.edu
[2] Vanderbilt University Medical Center, Nashville, TN 37215, USA

Abstract. Recent advances in deep learning opened a new window for achieving precise free water estimation (FWE) through single-shell diffusion-weighted imaging (DWI) data. Compared with traditional approaches, those methods mitigated biases from free water contamination without relying on multi-shell data. However, current single-shell-based FWE methods still suffer from the "case-by-case" deep learning design, struggling to generalize across the diverse acquisition schemes in a more clinically useful plug-and-play fashion. In this paper, we propose a novel token-aware single-shell FWE (Ts-FWE) method towards a "one-for-all" design, so that a single model is able to train and test on different shell configurations, even for unseen data. Specifically, Ts-FWE integrates a token that encapsulates the shell configuration with a Vision Transformer (ViT) backbone architecture. Moreover, 3D patches are employed to enhance the performance compared with traditional voxel-based modeling. Both cross-validation and external validation are performed through HCP young adults, aging and MASIVar datasets. The results demonstrated that the proposed method achieved superior performance, demonstrating a promising step towards building a more generalizable deep learning scheme for DWI.

Keywords: Diffusion MRI · Free water elimination · Vision transformer

1 Introduction

Diffusion magnetic resonance imaging (dMRI) is a noninvasive biomedical imaging technique to provide unique in vivo microstructural information, especially for the study of white matter structure and brain connectivity [3,9,25]. In the human brain, cerebrospinal fluid (CSF) confined to the ventricles and around the brain parenchyma as well as in lesion edema is considered free water. In dMRI, signals from free water in the CSF can 'contaminate' the image by the Free Water partial volume effects (PVE) that manifest at the boundary. The obtained MRI signal originates from both the CSF and as well as from the white matter partial, leading to potential misinterpretations [1]. These free water partial volume effects can strongly influence diffusion tractography.

The Free Water Elimination (FWE) model, as proposed by Pasternak et al. [21], aims to mitigate the adverse impact of CSF partial volume effects on diffusion measurements [12]. The initial model requires multiple b-values to distinguish the tissue types on the sub-voxel level [4,22]. More recent implementations were able to delineate fast diffusing components using single-shell data with the help of a-priori local spatial information [10]. Nonetheless, this traditional methodology becomes ill-posed in the absence of multi-shell constraints and certain underlying assumptions.

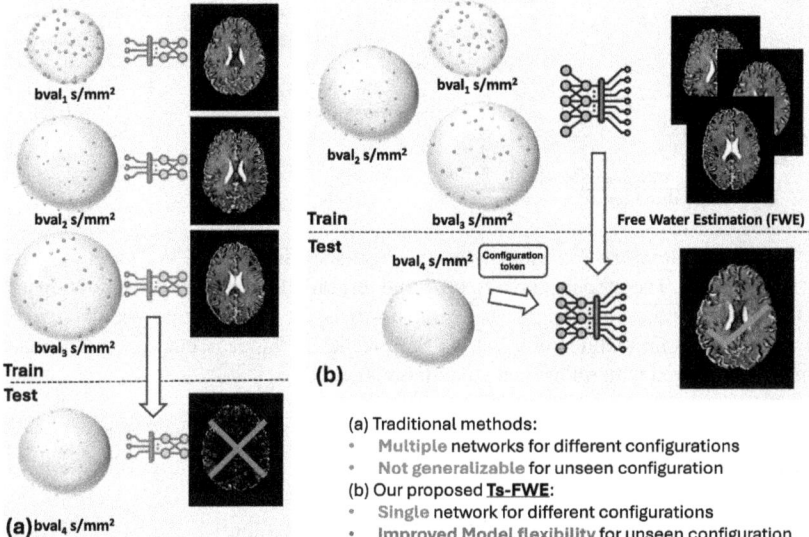

Fig. 1. Compare traditional methods and our proposed Ts-FWE method. For deep-learning-based diffusion MRI estimators, previous deep-learning approaches can achieve accurate model fitting. However, the "case-by-case" approach did not allow for variations in the input data acquisition scheme and therefore did not achieve a generalizable model for different data configurations. The proposed framework takes in a configuration token that encapsulates the shell configuration to improve accuracy over conventional deep estimators and improve model flexibility when dealing with unseen data configurations.

The traditional free water elimination methods have been effective across different b-values, but this success primarily applies to multi-shell diffusion MRI data. When it comes to single-shell data, the problem becomes ill-posed due to the lack of sufficient diffusion information, making these classical methods less reliable and more prone to inaccuracies. Recent breakthroughs in deep learning have ushered in a fresh opportunity for attaining accurate FWE using single-shell diffusion-weighted imaging (DWI) data. Yet, existing single-shell-based FWE techniques continue to grapple with a "case-by-case" deep learning approach [15–18,20,26], hindering their ability to generalize across various acquisition schemes and limiting their clinical utility in a more adaptable plug-and-play manner. Therefore, our goal is to build a unified model that can be

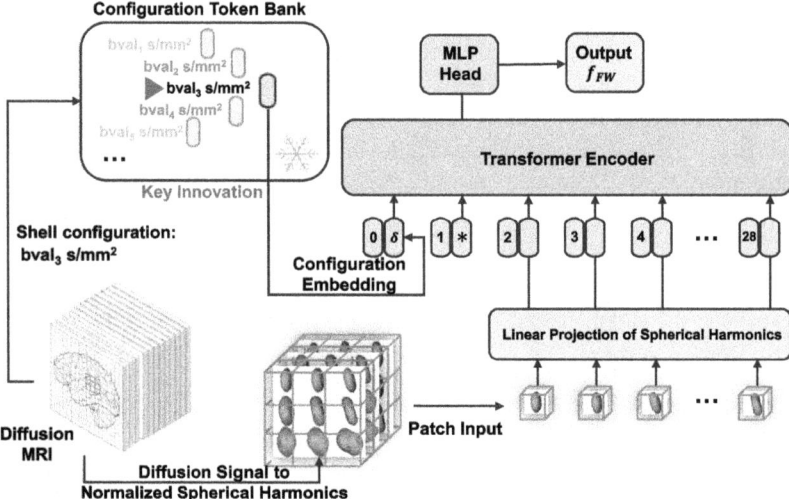

Fig. 2. Model Architecture. The token-aware single-shell FWE (Ts-FWE) method towards a "one-for-all" design. The model is able to train and test on different shell configurations. Specifically, Ts-FWE integrates a token that encapsulates the shell configuration with a Vision Transformer (ViT) backbone architecture. Taking 3D patches as input is employed to enhance the performance compared with traditional voxel-based modeling.

applied across various single-shell sequences, improving generalizability and robustness in practice.

In this study, we propose a novel token-aware single-shell FWE (Ts-FWE) method towards a "one-for-all" design, so that a single model can train and test on different shell configurations (Fig. 1). The contribution of this paper is four-fold:

- The proposed Ts-FWE method leverages the strengths of Vision Transformers (VIT) and introduces an innovative configuration token bank module derived from the shell configuration of single-shell data (as shown in Fig. 2).
- A single model is universally applicable to perform free water fraction estimation for heterogeneous single-shell diffusion MRI.
- 3D patches are employed to provide complete spatial information to enhance the performance compared with traditional voxel-based modeling.
- Experimental results demonstrate the superiority of our Ts-FWE framework over existing benchmark methods.

2 Related Work

Several DW-MRI methods have been developed that utilize the powerful data-driven capabilities of deep learning, yielding improved accuracy over conventional fitting when the acquisition scheme has a limited number of measurements. To search for a robust and precise way for dMRI free water estimation, Molina-Romero et al. [16] presented a deep learning framework in their study. This method employs a multi-layer perception (MLP) trained on a synthetic dataset of DWIs. This approach demonstrates the capability of deep learning models to interpret complex medical imaging data, providing a new avenue for more accurate and robust analysis in dMRI free water estimation, especially in clinical contexts where precise tissue characterization is crucial. However, their proposed deep learning frameworks are not generalizable to new acquisition schemes, as a case-by-case design (Fig. 1). The limitation of MLP (fixed input size) complicates applying a deep learning model to data acquired in different acquisition settings. Nath et al. [17, 18] have employed spherical harmonics coefficients as a representation for single shell dMRI, this solved the problem of fixed input size for MLP or convolutional neural networks. However, several studies [15, 20, 26] have shown the challenge of dealing with data from different shells in designing deep learning models for diffusion MRI lies in the inherent differences in the domains represented by each shell. These differences include variations in signal-to-noise ratio, contrast, and the specific diffusion properties they reveal. This heterogeneity makes it difficult to directly apply standard deep learning techniques that typically assume a uniform data domain. To improve deep learning generalizability on diffusion modeling, Yao et al. [26] employed a dynamic head (DH) design to handle the heterogeneous shell configuration problem. The dynamic headings offer a flexible way to handle diverse data types within a single model, adapting its behavior to best suit the input configuration. However, the network's generalizability is still limited since it could only be applied to data acquired within three b-values: 1000, 2000, and 3000 s/mm^2. When retraining the model on data that is acquired at a b-value beyond the training configuration set, the whole dynamic head and feature scaling module need to be redesigned.

3 Method

3.1 Backbone Design

Inspired by [2, 19], we utilize $3 \times 3 \times 3$ patches as input (shown in Fig. 2), as 3D patches provide more complete spatial information for deep learning networks compared with modeling each individual voxel independently. Each voxel is represented with 8_{th} order spherical harmonics (45 SH coefficients). The spherical harmonics coefficients of each voxel are projected into a higher-dimensional embedding space of size 2048. In this framework, each cubic patch forms a 'sentence' for the transformer model, with each voxel within the patch serving as a 'word' that contributes to the overall spatial context and representation. Additionally, layer normalization after this linear projection to normalize the embeddings. A dropout is applied to the embeddings to prevent overfitting by randomly zeroing some of the elements. Since transformers do not inherently understand the order of input tokens because the self-attention operation is permutation-invariant. To address this, positional embeddings are added to the patch embeddings to give the model a sense of order or position. The sequence of embeddings along with the positional embedding are passed through a series of 6 transformer layers. Each layer consists of a multi-head self-attention mechanism followed by a feedforward neural network. Residual connections and layer normalization are applied around both of these sub-layers. The self-attention mechanism is split into 8 heads, allowing the model to attend to different parts of the input differently, allowing the model to weigh the importance of different patches when considering a particular patch, and the feedforward networks apply further non-linear transformations. The MLP head (hidden dimensionality of 2048) in this ViT configuration acts as the final step in the regression pipeline, directly taking the processed representations from the transformer and outputting free water fraction predictions for the input central voxel of the input patches. Additional parameter adjustments were done during the estimation, we applied a mean diffusivity regularization threshold which is $2.7e-3 \, mm^2/s$. If the standard DTI diffusion tensor's mean diffusivity is almost near the free water diffusion value, the diffusion signal is assumed to be only free water diffusion (i.e. free water fraction will be set to 1 and tissue's diffusion parameters are set to zero).

3.2 Configuration Token Bank

Compared with the previous dynamic heading framework [26], instead of using dimensionally changeable one-hot vectors for configuration-aware encoding, we use dimensionally stable configuration-aware tokens joining the operation of embedding sequences to dynamically adjust its processing based on the acquisition parameters, thereby improving generalizability across different scanning protocols. We create a dynamic token bank during the initialization of the model using nn.init() function in PyTorch. Each token has the same dimensionality with patch embeddings and corresponds with a unique shell configuration. In this study, we included a larger set b-value of 1000, 1500, 2000, 2500, and $3000 \, s/mm^2$. This token bank was frozen and only served as an additional hint for the transformer encoder. The token-based head is more adaptable than the one-hot vector dynamic heading. When new data from unseen shells comes, the transformer-based model is more flexible to enlarge by simply enlarging the token bank rather than redesigning the architecture.

4 Data and Experiments

4.1 Data

Several datasets with distinct multi-shell configuration data and voxel sizes (suffer from different levels of partial volume effect) are chosen for developing the data-driven deep free water fraction estimator. We have chosen DW-MRI from the Human Connectome Project - Young Adult (HCP-ya) dataset [9,24], 45 subjects were used. The acquisitions had b-values of 1000, 2000, $3000 \, s/mm^2$ with 90 gradient directions on each shell. All HCP-ya dMRIs were distortion corrected with top-up and eddy [14]. 30 subjects were used as training data while 10 subjects were used as evaluation and 5 subjects as testing data.

The diffusion and T1-weighted MRI data of 50 unrelated healthy adults (age 50+) from the Lifespan HCP in Aging(HCP-A) [5] study was employed. Diffusion encoding was acquired with two shells of 1500 and $3000 \, s/mm^2$ (98–99 directions per shell) and with 28 b-value = $0 \, s/mm^2$ images interleaved. All datasets were acquired with two phase encoding directions: anterior to posterior (AP), and reversed phase encoding direction (PA). The data was preprocessed by the PreQual [6] default pipeline. 40 subjects were used to train the model while 5 subjects were used as evaluation and 5 subjects as testing data.

For the MASiVar dataset [7], five subjects were acquired on three different sites referred to as 'A', 'B', and 'C'. Structural T1 was acquired for all subjects at all sites. All in-vivo acquisitions were pre-processed with the PreQual pipeline [6]. The acquisitions at b-value of 1000, 1500, 2000, and 2500 s/mm^2 (each shell with 96 diffusion directions) were employed for the study. Three subjects from sets 'A' and 'B' were used as training data. While the other two form the test cohort. Scans from site 'C' were only used for unseen data evaluation.

The T1 of the same subject is extracted and registered to the mean b0 image and segmented for brain mask using SLANT [13]. All signal shell diffusion signals are transformed to normalized 8th order spherical harmonics signal ODF as a unified input for deep learning models, using spherical harmonics with the 'tournier07' basis [23]. The free water fractions are generated using the Free Water Elimination model from [12,21]. The free water DTI implementation detail was published in [11] and available in the DIPY library (version 1.70) with its default setting [8]. The free water fractions estimated from the full configuration of the multi-shell setting for each dataset are regarded as the silver standard in our study. A total of 5 shells (1000, 1500, 2000, 2500, and 3000 s/mm^2) are employed in this study. Due to data imbalance, the MASiVar dataset is oversampled during training.

4.2 Baseline Methods

We used VIT without configuration token hint and the dynamic head spherical convolution network($DH - F_{CSNN}$) from [26] as the baseline method for single shell dMRI free water fraction estimation. Both the Ts-FWE and baseline method used mean squared error(MSE)as the loss function. For the VIT model, the model was firstly only trained with single shell data from only one specific shell. Thus, we have eventually 5 independent models. Then the models were evaluated on the test cohort not only on the corresponding shell but also cross-evaluated on different configurations. Furthermore, another model is trained with the single-shell data regardless of the configuration. Without the configuration token, we assess whether the model could learn to adjust its representation learning by simply seeing the normalized spherical harmonics coefficients.

As for the benchmark deep learning framework, the dynamic deep learning framework from [26] is employed in this study. We firstly adjust the binary coding head of DH-F_{CSNN} from 3 to 5 and the feature scaling matrix is adjusted accordingly. As shown in Fig. 2. For the evaluation of deep learning-based voxel-wise modeling and patch-wise modeling, we keep the original voxel-wise design as one baseline method (shown as DH-F_{CSNN}). Additionally, the model is adjusted to take in $3 \times 3 \times 3$ patch input (shown as DH-$F_{CSNN}(3D)$). Each voxel is fed into the spherical convolution network individually. After feature extraction, three subsequent convolutional layers with skip connections serve as the critical components to process the 3D feature map. An MLP head (hidden dimensionality of 2048) takes the processed representations from

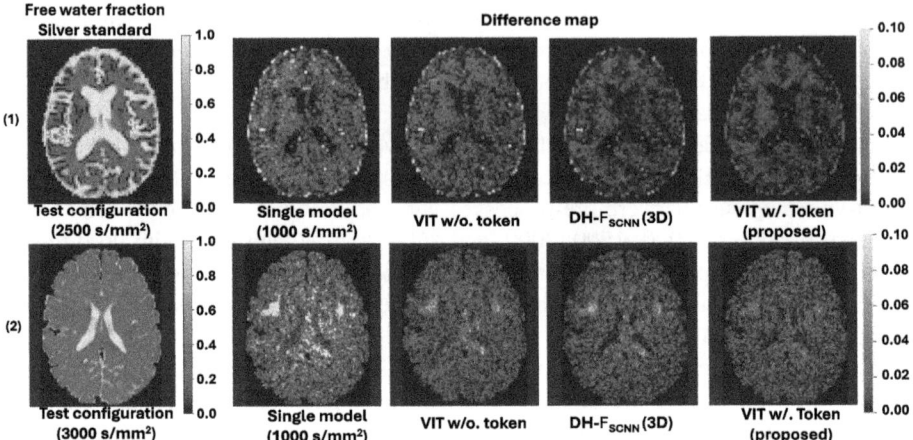

Fig. 3. Qualitative Result. Two subjects from different datasets have distinct voxel sizes that are visualized for assessment of free water estimation. Subject (1) is from MASIVar dataset (2.5 × 2.5 × 2.5 × 6.35) while subject (2) is from the HCP-Aging dataset (1.5 × 1.5 × 1.5 × 1). The silver standard free water map from multi-shell estimation is on the left panel, while the difference with the silver standard of the prediction from four models is visualized on the right panel, we set the scale to [0, 0.1], to highlight the relatively small difference.

the transformer and outputs free water fraction predictions for the input central voxel of the input patches.

For qualitative results, the free water fraction map is visualized of one subject from the MASIVar-'C' and another from the HCP-A, as shown in Fig. 3. The two subjects have distinct voxel sizes (MASIVar-'C': 1.5 × 1.5 × 1.5 × 1 and HCP-A: 2.5 × 2.5 × 2.5 × 6.35). For the MASIVar subject, we select 2500 s/mm^2 for testing and the single model was trained only on the configuration of 1000 s/mm^2 data. From the difference map, we can barely see the obvious different regions even though the colormap has been scaled to [0, 0.1]. The MSE loss is as low as 1.36E−04. The HCP-A subject, nonetheless, has a distinct voxel size, and the qualitative result shows a similar trend. The bigger voxel size leads to high differences around the brain area (boundaries between tissue and skull) and this phenomenon disappears in the HCP-A subject.

The MSE with predicted free water fractions from different methods and the silver standard is calculated and shown in Table 1. We first compared the performances between voxel-wise modeling and patch-wise modeling. The first two rows indicate patch-wise input has comparative improvement by providing the model with more spatial information with adjacent voxel. Furthermore, we evaluated different deep learning strategies' generalizability across different shell configurations. Our proposed method has very decent performances with the single model (when dealing with its corresponding configuration). However, the single model shows poor generalizability during cross-

Table 1. FW estimation on HCP-ya, HCP-A and MASIVar-'A, B'

Method	Testing configuration (s/mm^2)					Ave.	p-value
	1000	1500	2000	2500	3000		
DH-F_{CSNN}	2.46	2.79	2.40	3.08	2.72	2.69	$p<0.05$
DH-F_{CSNN} (3D)	2.34	2.71	2.31	3.17	2.64	2.63	$p<0.05$
Vit w/o. token ($1000s/mm^2$)	2.21	23.45	29.95	35.46	25.84	23.38	$p<0.05$
Vit w/o. token ($1500s/mm^2$)	20.02	2.43	28.68	35.81	25.61	22.51	$p<0.05$
Vit w/o. token ($2000s/mm^2$)	23.23	21.57	1.99	31.45	22.15	20.08	$p<0.05$
Vit w/o. token ($2500s/mm^2$)	26.35	25.29	25.33	2.72	23.21	20.58	$p<0.05$
Vit w/o. token ($3000s/mm^2$)	28.28	28.11	27.10	32.77	2.31	23.77	$p<0.05$
Vit w/o. token (all data)	2.53	2.55	2.67	2.89	2.58	2.64	$p<0.05$
Ts-FWE (Ours)	2.28	2.45	2.08	2.63	2.32	2.35	Ref.

All metric values are in units of 10^{-4}.
The top-2 performance is denoted as the red and blue mark, respectively. Statistical assessment is performed via the Wilcoxon signed-rank test.

Table 2. FW estimation on unseen site&subject:MASIVar-'C'

Method	Testing configuration (s/mm^2)				Ave.	p-value
	1000	1500	2000	2500		
DH-F_{CSNN}	2.93	2.83	2.77	3.09	2.90	$p<0.05$
DH-F_{CSNN} (3D)	2.75	2.74	2.55	2.88	2.73	$p<0.05$
Vit w/o. token ($1000s/mm^2$)	2.66	24.42	27.58	33.25	21.98	$p<0.05$
Vit w/o. token ($1500s/mm^2$)	24.58	2.48	27.66	30.25	21.24	$p<0.05$
Vit w/o. token ($2000s/mm^2$)	23.35	22.69	2.51	29.85	19.60	$p<0.05$
Vit w/o. token ($2500s/mm^2$)	30.35	32.51	28.33	2.69	23.47	$p<0.05$
Vit w/o. token (all data)	2.68	2.79	2.69	2.83	2.75	$p<0.05$
Ts-FWE (Ours)	2.61	2.52	2.43	2.56	2.53	Ref

All metric values are in units of 10^{-4}.
The top-2 performance is denoted as the red and blue mark, respectively. Statistical assessment is performed via the Wilcoxon signed-rank test.

Table 3. Model degradation assessment after configuration expansion on unseen site&subject:MASIVar-'C'

Model	Traning configuration (s/mm^2)	Testing configuration (1000 s/mm^2)
DH-F_{CSNN} (3D)	[1000, 2000, 3000]	2.60
DH-F_{CSNN} (3D)	[1000, 1500, 2000, 2500, 3000]	2.75
Ts-FWE (Ours)	[1000, 2000, 3000]	2.59
Ts-FWE (Ours)	[1000, 1500, 2000, 2500, 3000]	2.61

All metric values are in units of 10^{-4}.

evaluation. It is worth noting that as the b-value goes up, the proposed method outperforms the corresponding single model as the single model suffers from low SNR of diffusion signal. To further assess the model's generalizability, two test subjects from site 'C' in the MASIVar dataset are blinded from all the models (site-wise and subject-wise). The results are shown in Table 2. The proposed method outperformed other baseline methods in the configuration of 1000, 1500, and 2500 s/mm^2, which proves training on a variety of data lets our proposed method utilize data-driven techniques to infer plausible free-water volumes across different diffusion MRIs.

Furthermore, we assessed the model degradation while expanding the training configuration set and retrained the network. While expanding the training configuration, we reassigned the binary configuration coding for training data and additionally convolutional layers were introduced. For our method, we only need to adjust the token bank. As shown in Table 3, it illustrates that both the DH-FCSNN (3D) and Ts-FWE models show some degradation in performance when expanding the training configuration to include additional b-values. Notably, the DH-FCSNN (3D) model's metric worsens from 2.60 to 2.75, while the Ts-FWE model shows a much smaller increase from 2.57 to 2.61. This indicates that our Ts-FWE model exhibits significantly less degradation and flexibility for further development.

5 Conclusion

This study introduces the Ts-FWE framework for estimating free water fractions in brain diffusion MRI, with a particular focus on single-shell data. Our comprehensive evaluation demonstrates that the proposed method not only surpasses traditional deep learning-based approaches in terms of generalizability but also maintains high estimation accuracy. While tokenization and Vision Transformers (ViT) are not new techniques, their application to diffusion MRI estimation is novel and represents a significant advancement in the field. By incorporating a configuration token derived from the shell configuration of DWI data into a Vision Transformer model, this work establishes a new benchmark for the accurate and flexible estimation of free water fractions in single-shell diffusion MRI. Importantly, this research represents a crucial initial step towards developing more expansive and unified diffusion models that can adapt to a wide range of acquisition schemes, addressing the inherent variability in both b-vector and b-value.

References

1. Arezza, N.J., Santini, T., Omer, M., Baron, C.A.: Estimation of free water-corrected microscopic fractional anisotropy. Front. Neurosci. **17**, 1074730 (2023)
2. Bartlett, J., Davey, C., Johnston, L., Duan, J.: Recovering high-quality FODS from a reduced number of diffusion-weighted images using a model-driven deep learning architecture. arXiv preprint arXiv:2307.15273 (2023)
3. Basser, P.J., Mattiello, J., LeBihan, D.: Estimation of the effective self-diffusion tensor from the NMR spin echo. J. Magn. Reson., Ser. B **103**(3), 247–254 (1994)

4. Bergmann, Ø., Henriques, R., Westin, C.F., Pasternak, O.: Fast and accurate initialization of the free-water imaging model parameters from multi-shell diffusion MRI. NMR Biomed. **33**(3), e4219 (2020)
5. Bookheimer, S.Y., et al.: The lifespan human connectome project in aging: an overview. Neuroimage **185**, 335–348 (2019)
6. Cai, L.Y., et al.: Prequal: an automated pipeline for integrated preprocessing and quality assurance of diffusion weighted MRI images. Magn. Reson. Med. **86**(1), 456–470 (2021)
7. Cai, L.Y., et al.: Masivar: multisite, multiscanner, and multisubject acquisitions for studying variability in diffusion weighted MRI. Magn. Reson. Med. **86**(6), 3304–3320 (2021)
8. Garyfallidis, E., et al.: Dipy, a library for the analysis of diffusion MRI data. Front. Neuroinform. **8**, 8 (2014)
9. Glasser, M.F., et al.: The minimal preprocessing pipelines for the human connectome project. Neuroimage **80**, 105–124 (2013)
10. Golub, M., Neto Henriques, R., Gouveia Nunes, R.: Free-water DTI estimates from single b-value data might seem plausible but must be interpreted with care. Magn. Reson. Med. **85**(5), 2537–2551 (2021)
11. Henriques, R.N., Rokem, A., Garyfallidis, E., St-Jean, S., Peterson, E.T., Correia, M.M.: [re] optimization of a free water elimination two-compartment model for diffusion tensor imaging. BioRxiv 108795 (2017)
12. Hoy, A.R., Koay, C.G., Kecskemeti, S.R., Alexander, A.L.: Optimization of a free water elimination two-compartment model for diffusion tensor imaging. Neuroimage **103**, 323–333 (2014)
13. Huo, Y., et al.: 3D whole brain segmentation using spatially localized atlas network tiles. Neuroimage **194**, 105–119 (2019)
14. Jenkinson, M., Beckmann, C.F., Behrens, T.E., Woolrich, M.W., Smith, S.M.: FSL. Neuroimage **62**(2), 782–790 (2012)
15. Karimi, D., Vasung, L., Jaimes, C., Machado-Rivas, F., Warfield, S.K., Gholipour, A.: Learning to estimate the fiber orientation distribution function from diffusion-weighted MRI. Neuroimage **239**, 118316 (2021)
16. Molina-Romero, M., Gómez, P.A., Albarqouni, S., Sperl, J.I., Menzel, M.I., Menze, B.H.: Deep learning with synthetic data for free water elimination in diffusion MRI. In: Proceedings International Social Magnetic Resonance in Medicine (2018)
17. Nath, V., et al.: Enabling multi-shell b-value generalizability of data-driven diffusion models with deep SHORE. In: Shen, D., et al. (eds.) MICCAI 2019. LNCS, vol. 11766, pp. 573–581. Springer, Cham (2019). https://doi.org/10.1007/978-3-030-32248-9_64
18. Nath, V., et al.: Inter-scanner harmonization of high angular resolution DW-MRI using null space deep learning. In: Computational Diffusion MRI: International MICCAI Workshop, Granada, Spain, September 2018 22, pp. 193–201. Springer (2019)
19. Nath, V., et al.: Deep learning reveals untapped information for local white-matter fiber reconstruction in diffusion-weighted MRI. Magn. Reson. Imaging **62**, 220–227 (2019)
20. Ning, L., et al.: Cross-scanner and cross-protocol multi-shell diffusion MRI data harmonization: algorithms and results. Neuroimage **221**, 117128 (2020)
21. Pasternak, O., Sochen, N., Gur, Y., Intrator, N., Assaf, Y.: Free water elimination and mapping from diffusion MRI. Magnet. Reson. Med. Offic. J. Int. Soc. Magnet. Reson. Med. **62**(3), 717–730 (2009)
22. Scherrer, B., Warfield, S.K.: Why multiple b-values are required for multi-tensor models. Evaluation with a constrained log-euclidean model. In: 2010 IEEE International Symposium on Biomedical Imaging: From Nano to Macro, pp. 1389–1392. IEEE (2010)
23. Tournier, J.D., et al.: Mrtrix3: a fast, flexible and open software framework for medical image processing and visualisation. Neuroimage **202**, 116137 (2019)

24. Van Essen, D.C., et al.: The WU-MINN human connectome project: an overview. Neuroimage **80**, 62–79 (2013)
25. Van Essen, D.C., et al.: The human connectome project: a data acquisition perspective. Neuroimage **62**(4), 2222–2231 (2012)
26. Yao, T., et al.: A unified single-stage learning model for estimating fiber orientation distribution functions on heterogeneous multi-shell diffusion-weighted MRI. arXiv preprint arXiv:2303.16376 (2023)

Assessing Early Motor System Degeneration in the Spinal Cord of ALS Patients Using Diffusion MRI: An Exploratory Study

Alexandra Ford[1], Andrew W. Barritt[2,3], and Samira Bouyagoub[2(✉)]

[1] Brighton and Sussex Medical School, Falmer, Brighton, UK
[2] Clinical Imaging Sciences Centre, Department of Clinical Neuroscience, Brighton and Sussex Medical School, University of Sussex, Falmer, Brighton, UK
s.bouyagoub@bsms.ac.uk
[3] University Hospitals Sussex NHS Foundation Trust, Brighton, UK

Abstract. Amyotrophic Lateral Sclerosis (ALS) is a fatal neurodegenerative disease affecting primarily the motor system. The loss of upper motor neurons within the descending motor tracts, and lower motor neurons within the brain stem nuclei and spinal cord anterior horn gives rise to progressive muscle weakness. With the rapid course of the disease, there is a critical need for better biomarkers of disease to help understand the mechanisms of tissue damage in ALS and, ultimately, to improve diagnosis, prognosis and new treatments for patients. Diffusion Magnetic Resonance Imaging (dMRI) techniques, such as diffusion tensor imaging (DTI) and neurite orientation and dispersion density imaging (NODDI), provide a rich selection of parameters related to tissue microstructure. This project is a feasibility study aiming to evaluate dMRI metrics in the spinal cord as potential markers for ALS disease diagnosis. Multi-shell diffusion MRI data was collected from 11 recently diagnosed patients with ALS with upper limb symptoms, alongside 10 age-matched healthy controls, focusing on the cervical spinal cord region. Images underwent both DTI and NODDI fitting to produce fractional anisotropy (FA) and neurite orientation dispersion index (ODI) maps. Region-of-interest based statistical analyses were performed and results have shown significant FA reduction and ODI increase in ALS patients compared to controls, particularly in the white matter regions. These findings highlight the potential of diffusion MRI to uncover the nature of microstructural changes within the spinal cord during the early stages of ALS and opens the possibility to advance our understanding of the disease.

Keywords: NODDI · ALS · Spinal Cord MRI

1 Introduction

Amyotrophic Lateral Sclerosis (ALS) is an adult-onset neurodegenerative syndrome encompassing relentlessly progressive muscle weakness and varying degrees of cognitive impairment owing principally to loss of upper motor neuron (cortical pyramidal cells) and lower motor neurons within the brain stem nuclei and anterior horns of the spinal cord, but also to wider involvement of motor-associated networks [1]. Symptom presentation between patients is highly heterogeneous, with the average period between symptom onset and diagnosis approximately 12 months [2]. The delay in diagnosis reflects our incomplete understanding of the pathophysiology of ALS, where the exact mechanisms and sequence of degeneration remain subjects of debate [3,4]. Therefore, reliable and ideally non-invasive biological markers are required to enhance our understanding of these disease mechanisms. These markers could potentially enable not only the discovery of new targets for exploratory treatments, but also earlier clinical diagnosis before substantial cell loss occurs, and accurate prognosis.

There have been remarkable recent advancements in microstructural imaging, of the brain and the spinal cord, notably using diffusion MRI (dMRI). DMRI detects the movement of water on a microscopic level providing in-vivo insight to tissue microarchitecture [5]. Two of the most popular diffusion modelling techniques that have been previously explored in studying ALS include diffusion tensor imaging (DTI) [5] and Neurite Orientation Dispersion and Density Imaging (NODDI) [6]. DTI produces parameters such as Fractional Anisotropy (FA), which characterizes the anisotropy degree of water molecule diffusion within a tissue, and has been widely used in research studies that have shown that FA can predict the progression of ALS disease [7]. However, while FA is sensitive to detecting microstructural changes, it is not specific as to what causes those changes: whether it is loss of neurites or disorganization of neurites. NODDI is a more advanced technique that yields more specific parameters such as Orientation Dispersion Index (ODI) and Neurite Density Index (NDI) to assess the organization as well as the density of the fibers in tissue, respectively [6,8]. While there has been considerable research that explored both DTI and NODDI in understanding ALS, the research focused predominantly on the brain [9–11]. The spinal cord involvement is a prominent component of ALS but remains relatively less explored than the brain, and spinal cord studies continue to rely on the cords cross-sectional area (CSA) and DTI for investigating tissue damage [12,13]. Although NODDI has been successfully adapted and validated for studying the spinal cord [8,14], there have not yet been studies conducted on the human spinal cord using NODDI to study patients with ALS. Moreover, existing research literature appears to predominantly focus on patients during the advanced stages of the disease.

This study aims to investigate the early damage to central nervous system (CNS) in ALS patients using NODDI, focusing specifically on the cervical spinal cord. This is important not only to complement findings in the brain for a full comprehensive understanding of disease damage to the CNS, but also to evaluate whether NODDI can provide specific biomarkers to uncover the nature of the early microstructural changes within the spinal cord in ALS.

2 Methods

2.1 Participants

Eleven patients with ALS (six female; age range: 48–77 years, mean age 61 years, median age 60 years) with upper limb issues were recruited from the University Hospitals Sussex NHS Foundation Trust. Exclusion criteria included: Any contra-indications to MRI scanning or the inability to lie flat for 1 h due to severe illness or respiratory distress. Ten healthy, age-matched subjects (seven male; age range 33-79 years, mean age 56, median age 56 years) were also recruited as healthy controls. Patients clinical features such as the ALS Functional Rating Scale-Revised (ALSFRS-R) scores (median = 41/48) and disease duration (median = 17 months) were recorded at the time of recruitment. Site of onset of symptoms were recorded for each ALS patient: two subjects experienced bulbar onset whilst the remaining nine participants experienced initial limb weakness in either the upper or lower limbs. The median Kings Stage for our patient cohort was 2 [15], and median time from diagnosis to the MRI scan acquisition was 4 months. All participants had capacity and were able to provide informed consent. The study was approved by the Health Research Authority (Solihull Research Ethics Committee).

2.2 MRI Acquisition

Cervical spinal cord MRI data was acquired using a Siemens 3T Prisma scanner (Siemens, Erlangen, Germany) with a maximum gradient strength of 80 mT/m and a 64-channel head and neck coil, while the body coil was used for transmission. Only the elements surrounding the field of view (FOV) were used for image acquisition. Participants heads were padded to reduce movement during scanning. The participants heads were tilted slightly towards the neck to keep the cervical spine straight as possible throughout the duration of the scan whilst also keeping the participant comfortable.

Multi-shell diffusion MRI data was acquired, cardiac-triggered using a peripheral pulse oximeter with a trigger delay of 150 ms, to reduce the effect of cardiac motion on the image quality. The field of view was placed on the cervical cord area C4-T1 (see Fig. 1) with the following sequence parameters: TE = 86 ms, TR = 10beats (∼10000 ms); field of view = $86 \times 32 \, mm^2$; matrix size= 96×36; slice thickness = 5 mm and voxel size = $0.9 \times 0.9 \times 5 \, mm^3$. Two diffusion shells were collected: b = 1000 and 2850 s mm^{-2}, with 30 and 60 non-collinear diffusion directions, respectively. A further eleven non-diffusion weighted (b = 0) volumes were acquired. Additionally, structural T2-weighted images were acquired (TE = 120 ms; TR = 1500 ms; field of view = $256 \times 256 \, mm^2$; matrix size = 320×320; slice thickness = 0.8 mm). The T2 weighted MRI images were included to facilitate the spinal cord MRI data preprocessing, and to isolate the grey matter (GM) and white matter (WM) regions for region-of-interest (ROI) analysis. Illustrations of the grey and white matter structures and their location within a cross-section of the spinal cord are shown in Fig. 2 [16].

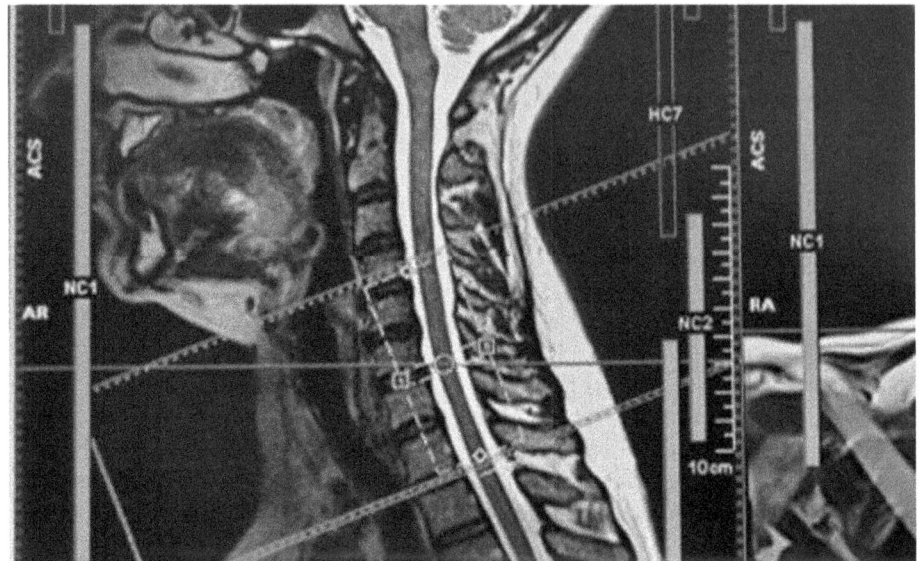

Fig. 1. Diffusion MRI acquisition field of view.

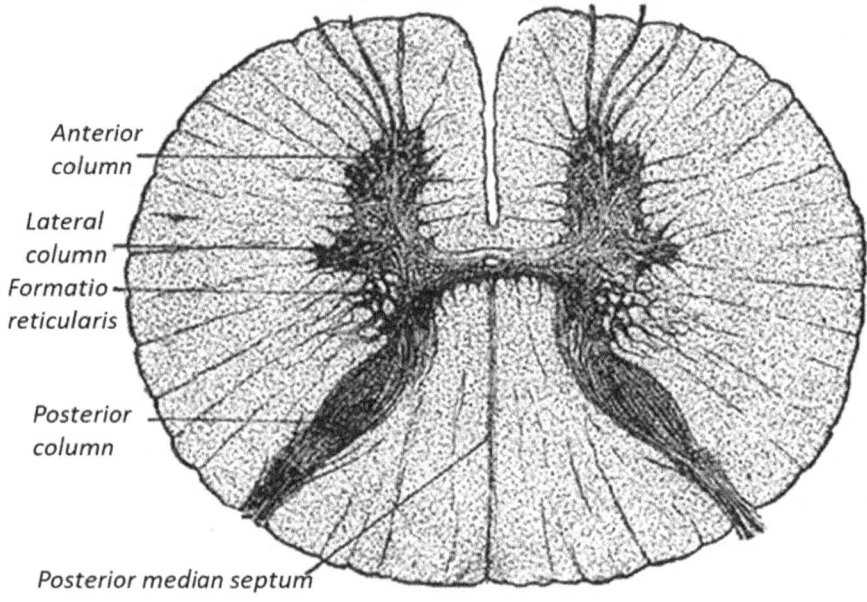

Fig. 2. Illustration of the gross anatomy of the spinal cord, highlighting the position of the grey matter and white matter. [16] (Color figure online)

2.3 Image Processing

The diffusion weighted MRI images (dMRI) were pre-processed using the spinal cord toolbox (SCT), see Fig. 3 for the full pipeline steps [17]. Raw images were first converted from the DICOM to NIfTI using dcm2niix (https://github.com/rordenlab/dcm2niix) [18]. Prior to motion correction, the images were cropped down to the region surrounding the cord to speed up subsequent processing steps and improve accuracy. For each subject, the mean image across all the dMRI volumes was calculated and then used to obtain a cord segmentation. A binary mask was then generated around the spinal cord which was subsequently used by the motion correction algorithm. The motion-corrected images were visually checked for quality.

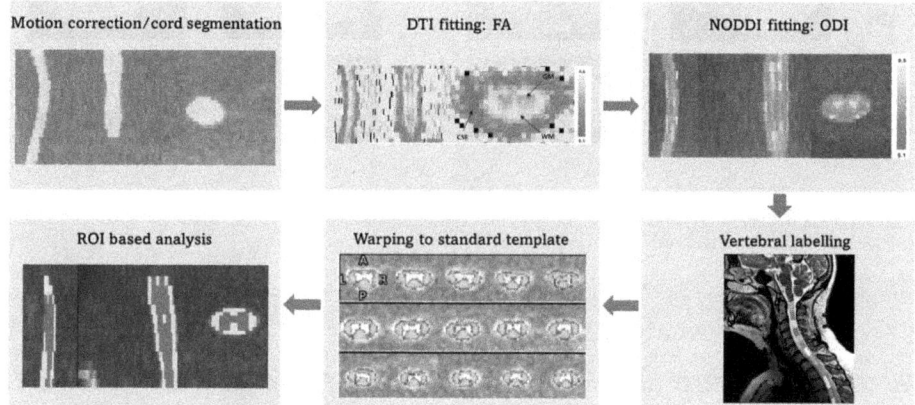

Fig. 3. Processing pipeline for spinal cord diffusion MRI data.

Once the preprocessing steps had been completed, DTI fitting was performed using the spinal cord toolbox and NODDI fitting was performed using the NODDI MATLAB Toolbox (http://mig.cs.ucl.ac.uk/index.php?n=Tutorial.NODDImatlab) run in Matlab 2023b (The MathWorks, Inc., Natick, MA). See Fig. 4 for an illustration of example slices from the ODI and FA maps.

Simultaneously, the T2 weighted images were used to perform vertebrae labelling. This is an important step to allow for registration to the spinal cord standard template - PAM50 template using vertebral matching [19]. PAM50 is an MRI template of the full spinal cord and brainstem that is available for both T1- and T2-weighted MRI images and includes atlases of both WM pathways and GM regions. Once the PAM50 template to native space warping parameters were calculated, they were utilized to transform the binary masks from the template space to the dMRI data native space. This step allows for extraction of metrics from the specific region of interest (C4-T1) as well as facilitating region of interest statistical analyses and between group comparison.

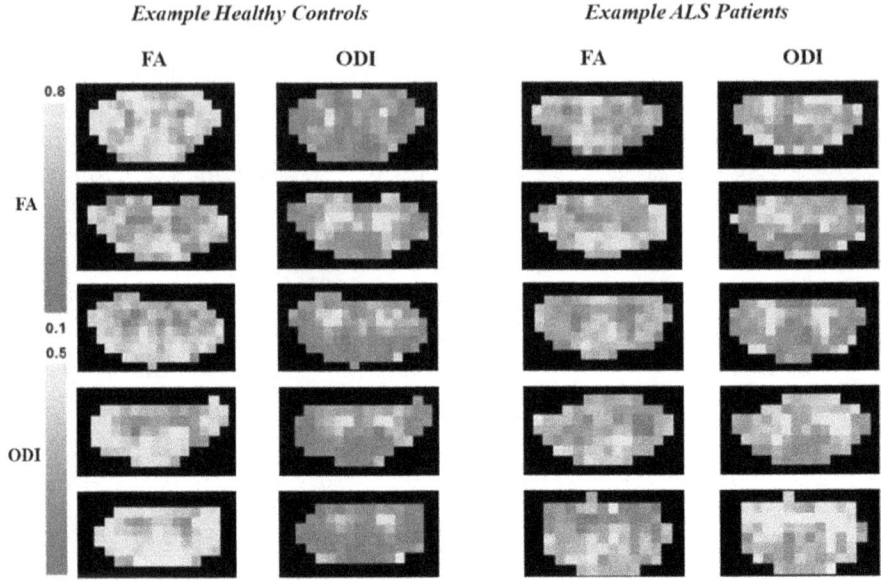

Fig. 4. Representative slices of ODI and FA maps in five different healthy controls compared to five different ALS patients.

2.4 Statistical Analysis

We used the PAM50 template masks to extract the mean values of the DTI and NODDI parameters within the WM region, GM region and at whole-cord level. FA and ODI parameter maps were selected for this statistical analysis to mitigate the impact of multiple comparisons due to a small cohort. Notably, our preliminary findings from a similar brain study [20] have indicated that among the diffusion metrics, FA and ODI can detect changes in the brain during the early stages of ALS.

To compare the NODDI and DTI-derived indices between the healthy controls and ALS patients, we conducted an Analysis of Covariance (ANCOVA). This analysis was performed separately for WM, GM and at whole-cord level, with age as a covariate. Research has highlighted ODI as particularly sensitive to aging [21]. Additionally, we explored correlations between FA and ODI and clinical features (disease duration and ALSFRS-R scores).

3 Results

Figure 4 offers a visual illustration of representative spinal cord FA and ODI slices in five different healthy subjects with a clearer contrast between white and grey matter, which disappears in slices corresponding to the same cord level in five different ALS patients. Group comparison results for ODI and FA between healthy controls and ALS patients in the WM, GM and at whole cord level are shown in Fig. 5.

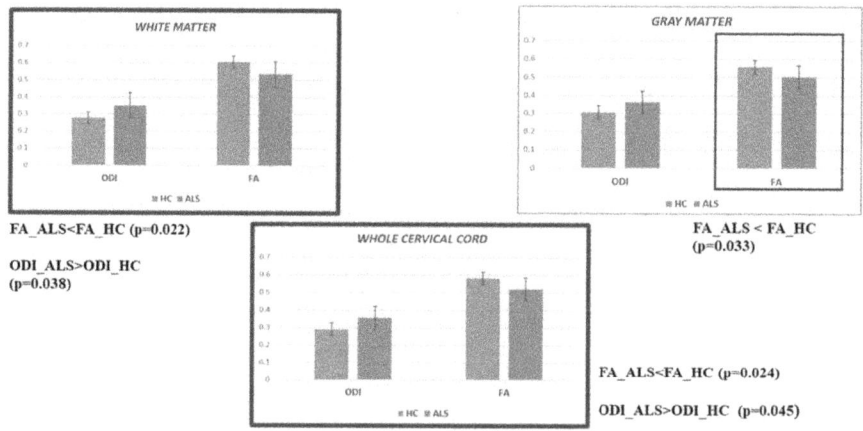

Fig. 5. Group analysis of ODI and FA maps between healthy controls and ALS patients in white matter (WM), grey matter (GM) and whole cervical cord. (Color figure online)

Within the whole cord, FA was reduced in the ALS group compared to healthy controls (p = 0.0242), and ODI was increased in the ALS group compared to the healthy controls (p = 0.0451). When examining the WM and GM separately, FA was reduced in both the WM (p = 0.0222) and GM (p = 0.0335) in the ALS group compared to the healthy controls. However, ODI was increased only in the WM region in the ALS group (p = 0.0382). There was no significant difference in ODI in the GM between the ALS group and healthy controls. There was no significant correlation between the clinical features (disease duration and ALSFRS-R scores) in the ALS group and the ODI/FA parameters.

4 Discussion

The aim of this study is to explore the cervical spinal cord using NODDI parameters for the first time in patients with ALS along with DTI parameters to better understand mechanisms of motor system damage in the early stages of the disease. The major findings of this study relate, firstly, to the significant decrease in FA in GM and WM and at whole-cord level and, secondly, to the significant increase in ODI within the WM and at whole-cord level of the cervical spine in the patients with ALS compared to healthy controls. These findings reflect the degeneration of structures within the cervical spine underlying the disease. Furthermore, they demonstrate the sensitivity of both DTI and NODDI to detect this damage at a median 4 months after diagnosis and with a relatively small number of subjects.

Reduced FA in WM has been demonstrated in multiple brain and spinal cord studies utilizing DTI techniques exploring ALS markers [9,22,23]. Our study examined a group of 11 patients at a median early-to-mid clinical stage of disease (median Kings Stage 2, median disease duration 17 months and median ALSFRS-R score 41/48) and showed reductions in FA in the WM, GM and at whole-cord level, which suggests global degeneration in the spinal cord at the levels C4-T1. Our findings support previous research showing reduced FA and cross-sectional area of the upper cervical part of the spinal cord in larger cohorts of ALS patients with greater disease burden [12,24].

Within the spinal cord, molecular diffusion is mostly anisotropic, restricted along the axons [13]. While FA is sensitive to changes to white matter integrity, NODDI offers an insight into the driver of this change in anisotropy, i.e. less fiber uniformity or loss of fibers [6,8]. Notably, the identified reductions in FA in our study were found to correlate with the increase in ODI scores, particularly within the white matter. Interestingly, increased ODI and extra-axonal space in the spinal cord has been shown in studies using G93A-SOD1 mice as an animal model for ALS [25]. Our findings suggest that early tissue damage in the cervical spinal cord detected through the decrease in FA is related to a disruption to the organization and architecture of the axons rather than necessarily neurite loss. Increased ODI in the WM also suggests that there may be increased extra-axonal space. There have been no studies to date that have used the NODDI technique in the cervical spine in humans with ALS, rendering our study findings very important because, by using ODI, we have underpinned the specific nature of the microstructural change detected by FA.

We found no significant correlation between FA or ODI and disease duration or the ALSFRS-R severity score, which is not dissimilar to some previous studies [12] albeit that others have shown relationships with ALSFRS-R [26,27]. However, previous research on the brain found a correlation between ODI and disease duration which suggests that there may be a loss of cortical dendrites alongside axonal degeneration [9]. The reason for no significant correlation with clinical factors in our data could be due to the small sample size of our patient cohort. While a reproducibility study on NODDI parameters in the brain has demonstrated that ODI can detect group differences even with small sample sizes [28], ALS is a highly heterogenous disease and statistical analyses involving clinical scores could benefit from sampling a larger more variable cohort.

This feasibility study is not without limitations, notably the small patient cohort and limiting our diffusion imaging to the cervical cord. However, our findings have demonstrated the potential for NODDI and DTI techniques to elucidate more understanding of the microstructural alterations within the spinal cord in ALS.

5 Conclusion

The results from this exploratory study offer a unique insight into the microstructural changes that occur in the cervical spinal cord at the early stages of the

ALS disease. By using NODDI on the spinal cord in ALS patients for the first time, the findings demonstrate the potential of NODDI to uncover the nature of the changes affecting the cord tissue. This opens the possibility to advance our understanding of the pathophysiology of ALS as well as contribute to the debate surrounding mechanism and sequence of damage within the motor system. Further extensive longitudinal studies with a larger cohort size are needed to evaluate NODDI measures as potential biomarker for ALS in order to aid diagnosis, prognosis, new hypotheses for potential therapeutic interventions and monitor response to future treatments within clinical trials, which is currently predominantly reliant on survival rates.

Acknowledgments. This study was funded by The Rising Star award from the University of Brighton. The authors would firstly like to thank all our patients and healthy controls for their participation in this research project. The authors would also like to thank all the radiographers at the Clinical Imaging Sciences Centre.

Disclosure of Interests. The authors have no competing interests to declare that are relevant to the content of this article.

References

1. Masrori, P., Van Damme, P.: Amyotrophic lateral sclerosis: a clinical review. Eur. J. Neurol. **27**(10), 1918–1929 (2020)
2. Richards, D., Morren, J.A., Pioro, E.P.: Time to diagnosis and factors affecting diagnostic delay in amyotrophic lateral sclerosis. J. Neurol. Sci. **417**, 117054 (2020)
3. Kiernan, M.C., et al.: Amyotrophic lateral sclerosis. Lancet **377**(9769), 942–55 (2011)
4. Zarei, S., et al.: A comprehensive review of amyotrophic lateral sclerosis. Surg. Neurol. Int. **6**, 171 (2015)
5. Alexander, A.L., Lee, J.E., Lazar, M., Field, A.S.: Diffusion tensor imaging of the brain. Neurotherapeutics **4**(3), 316–29 (2007)
6. Zhang, H., Schneider, T., Wheeler-Kingshott, C.A., Alexander, D.C.: NODDI: practical in vivo neurite orientation dispersion and density imaging of the human brain. Neuroimage **61**(4), 1000–16 (2012)
7. El Mendili, M.M., et al.: Multi-parametric spinal cord MRI as potential progression marker in amyotrophic lateral sclerosis. PLoS ONE **9**(4), e95516 (2014)
8. Grussu, F., et al.: Neurite orientation dispersion and density imaging of the healthy cervical spinal cord in vivo. Neuroimage **111**, 590–601 (2015)
9. Broad, R.J., et al.: Neurite orientation and dispersion density imaging (NODDI) detects cortical and corticospinal tract degeneration in ALS. J. Neurol. Neurosurg. Psychiatry **90**(4), 404–411 (2019)
10. Gabel, M.C., et al.: Evolution of white matter damage in amyotrophic lateral sclerosis. Ann. Clin. Transl. Neurol. **7**(5), 722–732 (2020)
11. Huang, N.X., et al.: Corticospinal fibers with different origins impair in amyotrophic lateral sclerosis: a neurite orientation dispersion and density imaging study. CNS Neurosci. Ther. **29**(11), 3406–3415 (2023)
12. Agosta, F., et al.: A longitudinal diffusion tensor MRI study of the cervical cord and brain in amyotrophic lateral sclerosis patients. J. Neurol. Neurosurg. Psychiatry **80**(1), 53–5 (2009)

13. Querin, G., et al.: Multimodal spinal cord MRI offers accurate diagnostic classification in ALS. J. Neurol. Neurosurg. Psychiatry **89**(11), 1220–1221 (2018)
14. Collorone, S., et al.: Reduced neurite density in the brain and cervical spinal cord in relapsing-remitting multiple sclerosis: a NODDI study. Mult. Scler. **26**(13), 1647–1657 (2020)
15. Roche, J.C., et al.: A proposed staging system for amyotrophic lateral sclerosis. Brain **135**, 847–852 (2012)
16. Gray, H., Standring, S., Ellis, H., Berkovitz, B.K.B.: Gray's Anatomy: The Anatomical Basis of Clinical Practice. 39th ed., p. 1627. Elsevier Churchill Livingstone, Edinburgh, New York (2005)
17. De Leener, B., et al.: SCT: Spinal Cord Toolbox, an open-source software for processing spinal cord MRI data. Neuroimage **145**(Pt A), 24–43 (2017)
18. Li, X., et al.: The first step for neuroimaging data analysis: DICOM to NIfTI conversion. J. Neurosci. Methods **264**, 47–56 (2016)
19. De Leener, B., et al.: PAM50: unbiased multimodal template of the brainstem and spinal cord aligned with the ICBM152 space. Neuroimage **165**, 170–179 (2018)
20. Bouyagoub, S., Lawden, E., Mamphey, D.F., Cercignani, M., Leigh, P.N., Barritt, A.: Longitudinal Study of Motor System Degeneration in the Brain of ALS Patients Using NODDI. [Manuscript in preparation] (2024)
21. Kodiweera, C., et al.: Age effects and sex differences in human brain white matter of young to middle-aged adults: a DTI, NODDI, and q-space study. Neuroimage **128**, 180–192 (2016)
22. Patzig, M., et al.: Measurement of structural integrity of the spinal cord in patients with amyotrophic lateral sclerosis using diffusion tensor magnetic resonance imaging. PLoS ONE **14**(10), e0224078 (2019)
23. Du, X.Q., et al.: Brain white matter abnormalities and correlation with severity in amyotrophic lateral sclerosis: An atlas-based diffusion tensor imaging study. J. Neurol. Sci. **405**, 116438 (2019)
24. Wimmer, T., et al.: The upper cervical spinal cord in ALS assessed by cross-sectional and longitudinal 3T MRI. Sci. Rep. **10**(1), 1783 (2020)
25. Gatto, R.G., et al.: Neurite orientation dispersion and density imaging can detect presymptomatic axonal degeneration in the spinal cord of ALS mice. Funct. Neurol. **33**(3), 155–163 (2018)
26. Valsasina, P., et al.: Diffusion anisotropy of the cervical cord is strictly associated with disability in amyotrophic lateral sclerosis. J. Neurol. Neurosurg. Psychiatry **78**(5), 480–484 (2007)
27. Cohen-Adad, J., et al.: Involvement of spinal sensory pathway in ALS and specificity of cord atrophy to lower motor neuron degeneration. Amyotrophic Lateral Sclerosis and Frontotemporal Degeneration **14**(1), 30–38 (2013)
28. Bouyagoub, S., Dowell, N.G., Gabel, M., Cercignani, M.: Comparing multiband and singleband EPI in NODDI at 3 T: what are the implications for reproducibility and study sample sizes? MAGMA **34**(4), 499–511 (2021)

RobNODDI: Robust NODDI Parameter Estimation with Adaptive Sampling Under Continuous Representation

Taohui Xiao[1,4], Jian Cheng[2,3], Wenxin Fan[4], Jing Yang[4], Cheng Li[4], Enqing Dong[1(✉)], and Shanshan Wang[4,5(✉)]

[1] School of Mechanical, Electrical and Information Engineering, Shandong University, Weihai 264209, China
enqdong@sdu.edu.cn
[2] State Key Laboratory of Software Development Environment, Beihang University, Beijing, China
[3] Key Laboratory of Data Science and Intelligent Computing, Institute of International Innovation, Beihang University, Hangzhou, Zhejiang, China
[4] Paul C. Lauterbur Research Center for Biomedical Imaging, Shenzhen Institute of Advanced Technology, Chinese Academy of Sciences, Shenzhen, Guangdong, China
ss.wang@siat.ac.cn
[5] Peng Cheng Laboratory, Shenzhen, Guangdong, China

Abstract. Neurite Orientation Dispersion and Density Imaging (NODDI) is an important imaging technology used to evaluate the microstructure of brain tissue, which is of great significance for the discovery and treatment of various neurological diseases. Current deep learning-based methods perform parameter estimation through diffusion magnetic resonance imaging (dMRI) with a small number of diffusion gradients. These methods speed up parameter estimation and improve accuracy. However, the diffusion directions used by most existing deep learning models during testing needs to be strictly consistent with the diffusion directions during training. This results in poor generalization and robustness of deep learning models in dMRI parameter estimation. In this work, we first verify that the parameter estimation performance of current mainstream methods will significantly decrease when the testing diffusion directions and the training diffusion directions are inconsistent. A robust NODDI parameter estimation method with adaptive sampling under continuous representation (RobNODDI) is proposed. Furthermore, long short-term memory (LSTM) units and fully connected layers are selected to learn continuous representation signals. To this end, we use a total of 100 subjects to conduct experiments based on the Human Connectome Project (HCP) dataset, of which 60 are used for training, 20 are used for validation, and 20 are used for testing. The test results indicate that RobNODDI improves the generalization performance and robustness of the deep learning model, enhancing the stability and flexibility of deep learning NODDI parameter estimatimation applications.

Keywords: Robustness · Diffusion MRI · Adaptive sampling · Continuous representation · NODDI parameter estimation

1 Introduction

Diffusion magnetic resonance imaging (dMRI) provides a unique tool for non-invasive assessment of tissue microstructure [1]. Neurite Orientation Dispersion and Density Imaging (NODDI) is a popular physiological component-based microstructural model in dMRI [2]. NODDI-derived measures can reflect changes in brain microstructural properties across a variety of neurological and psychiatric disorders, as well as brain development, maturation, and aging across the lifespan, including from neonatal to adolescence and adulthood [3–7].

Advanced dMRI models are highly nonlinear and composed of multiple compartments, so they often require dense sampling in q-space to better estimate parameters. Dense sampling in q-space requires the acquisition of many diffusion-weighted images (DWI) with different diffusion directions and b-value, which is very time-consuming and prone to motion artifacts [8]. In the current parameter estimation of advanced dMRI models, optimization-based methods such as nonlinear least squares (NLLS) method, the Markov chain Monte Carlo (MCMC) method, and Bayesian method are easy to produce estimation errors [1]. Moreover, deep learning-based methods have been widely used in microstructure estimation research [8–20]. Golkov et al. [9] used multilayer perceptron (MLP) for the first time to estimate diffusion kurtosis and microstructural parameters. Next various deep learning networks were used for dMRI model estimation. Chen et al. [10,17–19] proposed graph neural network (GNN) based structure to estimate the NODDI model parameters. Ye et al. [8,11–13] proposed a series of deep learning models to estimate NODDI model parameters. In addition, there are also studies [14–16] linking spherical harmonic (SH) coefficients to dMRI, among which Vishwesh et al. [16] directly used SH coefficients to estimate the microstructure. However, this study used images in the full gradient direction of single-shell, did not perform downsampling operations on single-shell, and did not fully exploit the multi-shell signal of dMRI. In summary, deep learning microstructure parameter estimation has achieved good results, speeding up parameter estimation and improving accuracy. However, the current testing process of deep learning models in microstructural parameter estimation needs to be highly consistent with training, that is, fixed DWI gradient direction information needs to be used. This results in poor generalization and clinical applicability of the deep learning model in dMRI microstructural parameter estimation, which does not meet the actual clinical needs.

To overcome these problems, in this article, we propose a robust adaptive sampling NODDI parameter estimation method under continuous representation (RobNODDI). We cut the original DWI into four-dimensional patches, then perform adaptive sampling, and then convert it into a continuous representation signal through SH fitting. Adaptive sampling can ensure that as much data as possible is involved in training, and the model can mine more useful information. Continuous representation helps the model be more flexible when testing diffusion directions and the training diffusion directions are inconsistent. These operations help our model achieve better results in random sampling tests. It is worth noting that our method has no special requirements for model architecture and is theoretically suitable for most current deep learning models [21–23]. Therefore, we selected the advanced MESC-SD as our basic architecture [12].

In this study, our method was verified in the NODDI model. The dataset is the public the Human Connectome Project (HCP) dataset [24]. The results show that the proposed method can greatly improve the generalization and robustness of existing deep learning models, and has better clinical applicability.

2 Method

RobNODDI combines adaptive sampling and continuous representation. The purpose of adaptive sampling is to improve the network's adaptability to different sampling data and fully mine and use the information in DWI. Continuous representation can make subsequent testing tasks more flexible by converting DWI into SH coefficients and allowing the model to directly learn this continuous representation. These strategies ensure that the diffusion directions of RobNODDI during testing does not need to be consistent with the diffusion directions during training, can effectively estimate parameter results, and significantly improve the generalization [25] and robustness of the deep learning model.

2.1 Overall Architecture

As shown in Fig. 1, our method mainly has two core points: adaptive sampling and continuous representation. We first obtain DWI patches of a total of D diffusion directions for any two shells (bi and bj). During the training stage, we divide multi-shell into two independent shells and perform adaptive sampling respectively. A single shell can be sampled as w × w × w × N, N represents the number of diffusion directions for adaptive sampling, which can be set between 20 and 60. Then the SH coefficients are calculated through linear least squares on the sampled patches, which is a continuously represented signal. Then we concatenate the SH coefficients of the two shells as input to the model for training. The size of the output parameter patch during training is generally set to be smaller than the input patch, and we set it to (w-2) × (w-2) × (w-2) × 3. In the testing stage, we can accurately estimate the NODDI parameters by inputting DWI patches of two shells with sizes w × w × w × S1 and w × w × w × S2 respectively. Among them, S1 and S2 can be different from N in training, so RobNODDI will be more flexible and robust in clinical applications.

2.2 Continuous Representation

The SH function is a special function defined on the sphere and has orthogonal completeness and continuity. It can be used as a set of orthogonal complete bases for functions on the sphere, and is used to expand and approximate functions

156   T. Xiao et al.

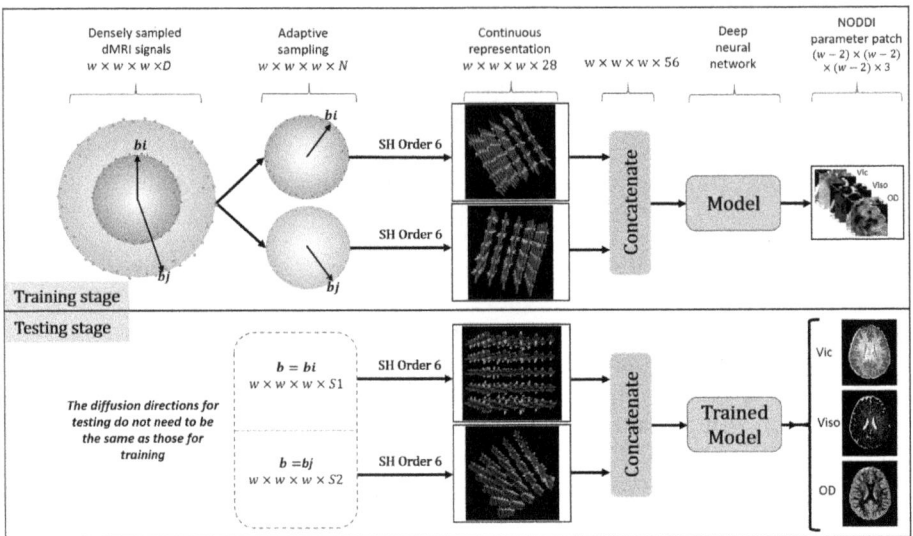

Fig. 1. Overview of RobNODDI. It contains training stage and testing stage. RobNODDI performs adaptive sampling and SH fitting on the DWI patches, and then concatenates the SH coefficients into the model. Note that both adaptive sampling and SH fitting are included in the training stage, and only SH fitting is included in the testing stage.

on the sphere. Therefore, it is often used in dMRI [14–16,26,27]. The expansion formula of dMRI signal on the sphere is shown in formula (1).

$$D(\theta, \varphi) = \sum_{l=0}^{\infty}\sum_{m=-l}^{l} \hat{C}_l^m Y_l^m(\theta, \varphi) \tag{1}$$

Here, $D(\theta, \varphi)$ represents the normalized dMRI signal, $\hat{C}_l^m$ represents the SH coefficient of dMRI, $Y_l^m(\theta, \varphi)$ represents the SH basis. l and m represent SH order and degree respectively. In dMRI, the direction of the diffusion gradient is described by θ, φ. Where θ and φ respectively represent two angles in the spherical coordinate system, namely the polar angle and the azimuth angle. When the basis is formed, the SH coefficients $\hat{C}_l^m$ can be solved by regularized linear least squares fitting [14]. In practical applications, with the original DWI data and corresponding direction coordinate information, we can obtain the corresponding the SH coefficients by solving formula (1).

2.3 Adaptive Sampling

When sampling DWI signal in dMRI microstructure parameter estimation, uniform sampling is generally performed in q-space [11,26–28], which ensures a more balanced use of the information in DWI. But for deep learning models, the sampling of the model during testing must be consistent with the training, which is inflexible and leads to a greatly reduced generalization and robustness of the model. Therefore, in this study, RobNODDI utilizes a combination of adaptive sampling and continuous representation. Specifically, each DWI patch used for training is randomly sampled, and the sampling pattern varies across different training epochs. Then, we perform Spherical Harmonics (SH) fitting on the adaptively sampled DWI patch, and finally input the processed patch into the network for training. In actual implementation, We first cut the DWI of the original two b values into w × w × w × D, where the fourth dimension D represents all directions of the two arbitrary b values. Then we randomly sample the patches of the two shells separately to obtain patches of size w × w × w × N, where N can be set between 20 and 60. We perform random sampling for each patch to enable the network to achieve adaptive sampling. After training on patches containing N diffusion directions, our model can be tested in multi-shell using S directions per shell that are different from the trained directions. In theory, it has better robustness than existing deep learning methods.

2.4 Network Construction

Our proposed method is independent of the network architecture used. Therefore, in this study, we utilize the MESC-SD developed for NODDI used in the study [12] as the basic architecture. The network architecture adopts SD-LSTM in the first step, which is constructed by unfolding an iterative process for solving the sparse reconstruction problem, and in the second stage by adding a fully connected layer to calculate the mapping of SH coefficient representations to NODDI parameter.

3 Experiments

In this section, we evaluate the proposed method to estimate NODDI parameters, including intracellular volume fraction (V_{icvf}), isotropic volume fraction (V_{iso}), and directional dispersion (OD) [2,12]. The network generalization performance is compared with deep learning methods such as q-DL [9], GCNN [10] and MESC-SD [12] through SS and RS tests. In particular, we use SS to indicate that the same sampling scheme is used during both testing and training. On the other hand, RS is used to indicate that random sampling is employed during testing, where the diffusion directions in the testing set are different from those used in the training set.

3.1 Implementation Details

We used 6-order SH (28 coefficients) in the HCP dataset, with N = 30 diffusion directions for each shell, and select two shells of b = 1000 s/mm^2 and b = 2000 s/mm^2. We set w = 5, which means cutting the patch into a size of 5 × 5 × 5 × 180, and the corresponding patch size of the NODDI parameter is 3 × 3 × 3 × 3. Additionally, mean square error (MSE) is used as the loss function and Adam is used as the optimizer. The initial learning rate and update method of different experiments are different. The initial learning rate is mainly set to 0.0005 and 0.0001. The training epochs are 30 or 50, and the batchsize of all experiments is set to 128. All models are applied in pytorch2.0 and trained on a server equipped with TITAN Xp GPU.

3.2 Dataset and Evaluation Metrics

Dataset. We selected data from HCP dataset [21] to conduct experiments and test the results. The data set contains 3 b values (b = 1000, 2000, 3000 s/mm^2), 90 diffusion directions for each b value, and an additional 18 non-diffusion weighted volumes. We randomly selected 100 adult subjects and used 60% of them for training, 20% for validation, and the remaining 20% for testing. We obtain the gold standard by fitting all spherical shells with all gradient directions through AMICO [29].

Evaluation Metrics. We use the root mean square error (MSE), peak signal-to-noise ratio (PSNR), and structural similarity index measure (SSIM) to evaluate the quality of the predicted NODDI-derived indices.

3.3 Results

The whole experiment includes: Testing the poor generalization of current deep learning methods in parameter estimation under different diffusion directions. Evaluating whether fitting DWI with SH as input can help improve the generalization of existing methods. Demonstrating that the proposed RobNODDI method achieves the best robustness while ensuring good performance.

Table 1. Average quantitative indicators of NODDI parameters for SS and RS testing using different methods on the q-space data with 30 diffusion directions per shell (1000, 2000 s/mm^2).

Method	Sampling in testing	MSE	PSNR	SSIM
q-DL [9]	SS	0.00463	23.41919	0.92024
	RS	0.00851	20.74516	0.88076
GCNN [10]	SS	0.00405	23.97635	0.92708
	RS	0.01268	19.00210	0.86630
MESC-SD [12]	SS	0.00219	26.64520	0.95775
	RS	0.00842	20.79057	0.89852
MESC-SD with SH	SS	0.00226	26.49617	0.95626
	RS	0.00263	25.82884	0.94980
RobNODDI	SS	**0.00211**	**26.80304**	**0.95901**
	RS	**0.00211**	26.79317	0.95884

SS and RS Tests of Existing Deep Learning Methods. Table 1 shows the quantitative metrics of three deep learning methods under SS and RS testing. It can be observed that the existing deep learning methods suffer a significant performance drop when the diffusion direction changes, making it difficult to estimate the parameters accurately. The visual comparison in Fig. 2 also demonstrates that the existing deep learning methods exhibit a clear performance degradation when the testing diffusion direction is altered. In other words, most existing deep learning methods have poor generalization to changes in dMRI diffusion direction.

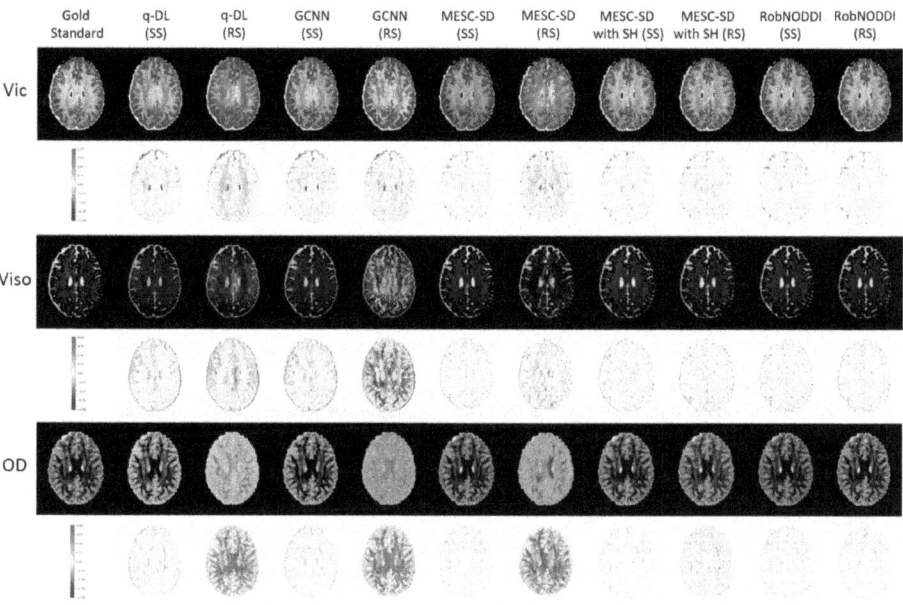

Fig. 2. Qualitative comparison of NODDI parameter for SS and RS testing using different methods with 30 diffusion directions per shell (1000, 2000 s/mm^2)

SS and RS Testing Using SH Coefficients as Input. The input for training here is the SH coefficient of 30 diffusion directions with fixed uniform sampling per shell. The test is also divided into two types: SS and RS. The results are shown in Fig. 2 and Table 1 corresponding to MESC-SD with SH (SS) and MESC-SD with SH (RS). It can be seen that using SH as input to estimate microstructure parameters is effective, and the results are very close to using DWI as input. When the test direction is RS, the performance also decreases; however, compared to the original MESC-SD, the RS test shows much better results. This indicates that using continuous representation of DWI as input improves the model's generalization performance in dMRI microstructure parameter estimation, yet it still lags behind the SS test. Specifically, for Vic and OD, the visual results and error maps in Fig. 2 still show some discrepancy compared to the SS testing results.

SS and RS Test of the Proposed RobNODDI Method. This part introduces our proposed RobNODDI method. As shown in Fig. 1, our input is the DWI patch, and we incorporate adaptive sampling and SH coefficient fitting into the network training. During testing, we also conducted SS and RS evaluations for RobNODDI, where the SS testing uses the same sampling scheme as the comparative methods. The corresponding results are shown in Fig. 2 and Table 1. It can be observed that when the testing direction changes, the results estimated by our method remain highly consistent, with almost no visible differences, and the quantitative metrics are very close as well. Compared to the original MESC-SD method, our approach achieves significantly improved generalization. Moreover, compared to MESC-SD with SH, the proposed method further enhances both performance and generalization. Therefore, our method not only achieves more accurate parameter estimation, but also demonstrates strong generalization and robustness.

3.4 Ablation Study

Table 2. RobNODDI performs RS test of NODDI parameters on q-space data with different number of diffusion directions (1000, 2000 s/mm^2)

Method	Sampling in testing	Number of directions	MSE	PSNR	SSIM
RobNODDI	RS	20/20	0.00231	26.40777	0.95512
		25/25	0.00217	26.67221	0.95759
		30/30	0.00211	26.80304	0.95901
		35/35	0.00208	26.85303	0.95952
		40/40	0.00208	26.87385	0.95989
		16/29	0.00226	26.50507	0.95590
		21/28	0.00219	26.64630	0.95730
		26/23	0.00219	26.64381	0.95732

To further validate the robustness of the proposed method, we conducted ablation experiments on RobNODDI by randomly sampling DWI with varying diffusion directions and numbers. This included tests with consistent numbers of diffusion directions for two b values: 40, 50, 70, and 80 directions (20, 25, 35 and 40 for b = 1000 s/mm^2 and b = 2000 s/mm^2). And tests with different numbers of diffusion directions for two b values: 45, 49, and 49 directions (16, 21 and 26 for b = 1000 s/mm^2; 29, 28 and 23 for b = 2000 s/mm^2). The ablation results are presented in Table 2, highlighting the robustness of RobNODDI.

It is evident that performance slightly improves with an increasing number of diffusion directions used in testing. Moreover, our method consistently maintains stable performance and high flexibility across different numbers of DWI signals used for testing.

4 Conclusion and Future Work

In this work, we propose RobNODDI, which achieves robust NODDI parameter evaluation by combining adaptive sampling and continuous representation. By comparing with the existing deep learning microstructure estimation methods, it is shown that converting dMRI images into SH coefficients helps to improve the generalization of the original method for diffusion directions. The proposed RobNODDI can further improve the robustness and stability of the deep learning model in dMRI microstructure parameter estimation, making the deep learning model more flexible and universal in dMRI clinical applications. Moreover, the model architecture in the method does not rely on a specific network structure. In our future work, we will explore more advanced methods, such as [15,30], and further optimize the proposed approach.

Acknowledgement. This research was partly supported by the National Natural Science Foundation of China (62222118, U22A2040, 62171261, 81671848, 81371635), Guangdong Provincial Key Laboratory of Artificial Intelligence in Medical Image Analysis and Application (2022B1212010011), Shenzhen Science and Technology Program (RCYX20210706092104034, JCYJ20220531100213029), Key Laboratory for Magnetic Resonance and Multimodality Imaging of Guangdong Province (2023B1212060052), Youth Innovation Promotion Association CAS, National Key R&D Program of China (2023YFA1011400), Fundamental Research Funds for the Central Universities (China), and Innovation Ability Improvement Project of Science and Technology Small and Medium-sized Enterprises of Shandong Province (2021TSGC1028).

References

1. Johansen-Berg, H., Behrens, T. E.: Diffusion MRI: From Quantitative Measurement to in Vivo Neuroanatomy. Academic Press (2016)
2. Zhang, H., Schneider, T., Wheeler-Kingshott, C.A., et al.: NODDI: practical in vivo neurite orientation dispersion and density imaging of the human brain. Neuroimage **61**(4), 1000–1016 (2012)
3. Greenspan, H., Van Ginneken, B., Summers, R.M.: Guest editorial deep learning in medical imaging: overview and future promise of an exciting new technique. IEEE Trans. Med. Imaging **35**(5), 1153–1159 (2016)
4. Fu, X., Shrestha, S., Sun, M., et al.: Microstructural white matter alterations in mild cognitive impairment and Alzheimer's disease: study based on neurite orientation dispersion and density imaging (NODDI). Clin. Neuroradiol. **30**, 569–579 (2020)
5. Kunz, N., Zhang, H., Vasung, L., et al.: Assessing white matter microstructure of the newborn with multi-shell diffusion MRI and biophysical compartment models. Neuroimage **96**, 288–299 (2012)

6. Mah, A., Geeraert, B., Lebel, C.: Detailing neuroanatomical development in late childhood and early adolescence using NODDI. PLoS ONE **12**(8), e0182340 (2017)
7. Billiet, T., Vandenbulcke, M., Mädler, B., et al.: Age-related microstructural differences quantified using myelin water imaging and advanced diffusion MRI. Neurobiol. Aging **36**(6), 2107–2121 (2015)
8. Zheng, T., Zheng, W., Sun, Y., et al.: An Adaptive Network with Extragradient for Dif-fusion MRI-Based Microstructure Estimation. In: Wang L. W., Dou Q., Fletcher P. T., Speidel S., Li S. (eds.) MICCAI 2022, LNCS, vol. 13431, pp.153–162. Springer, Cham (2022). https://doi.org/10.1007/978-3-031-16431-6_15
9. Golkov, V., Dosovitskiy, A., Sperl, J.I., et al.: Q-space deep learning: twelve-fold shorter and model-free diffusion MRI scans. IEEE Trans. Med. Imaging **35**(5), 1344–1351 (2016)
10. Chen, G., Hong, Y., Zhang, Y., et al.: Estimating tissue microstructure with undersampled diffusion data via graph convolutional neural networks. In: Martel, A.L., et al. (eds.) MICCAI 2020, LNCS, vol. 12267, pp. 280–290. Springer, Cham (2020). https://doi.org/10.1007/978-3-030-59728-3_28
11. Ye, C., Li, X., Chen, J.: A deep network for tissue microstructure estimation using modi-fied LSTM units. Med. Image Anal. **55**, 49–64 (2019)
12. Ye, C., Li, Y., Zeng, X.: An improved deep network for tissue microstructure estimation with uncertainty quantification. Med. Image Anal. **61**, 101650 (2020)
13. Zheng, T., Yan, G., Li, H., et al.: A microstructure estimation Transformer inspired by sparse representation for diffusion MRI. Med. Image Anal. **86**, 102788 (2023)
14. Koppers, S., Bloy, L., Berman, J.I., Tax, C.M.W., Edgar, J.C., Merhof, D.: Spherical harmonic residual network for diffusion signal harmonization. In: Bonet-Carne, E., Grussu, F., Ning, L., Sepehrband, F., Tax, C. (eds.) Computational Diffusion MRI. MICCAI Workshop 2019. Mathematics and Visualization. pp.173–182. Springer, Cham (2019). https://doi.org/10.1007/978-3-030-05831-9_14
15. Sedlar, S., Alimi, A., Papadopoulo, T., Deriche, R., Deslauriers-Gauthier, S.: A Spherical convolutional neural network for white matter structure imaging via dMRI. In: de Bruijne, M., et al. (eds.) MICCAI 2021. LNCS, vol.12903, pp.529–539. Springer, Cham (2021). https://doi.org/10.1007/978-3-030-87199-4_50
16. Nath, V., et al.: DW-MRI microstructure model of models captured via single-shell bottleneck deep learning. In: Gyori, N., Hutter, J., Nath, V., Palombo, M., Pizzolato, M., Zhang, F. (eds.) Computational Diffusion MRI. MICCAI Workshop 2019. Mathematics and Visualization, pp.147–157. Springer, Cham (2021). https://doi.org/10.1007/978-3-030-73018-5_12
17. Yang, J., et al.: Towards accurate microstructure estimation via 3D hybrid graph transformer. In: Greenspan, H., et al (eds.) MICCAI 2023. LNCS, vol. 14227, pp.25–34. Springer, Cham (2023). https://doi.org/10.1007/978-3-031-43993-3_3
18. Chen, G., et al.: Hybrid graph transformer for tissue microstructure estimation with undersampled diffusion MRI data. In: Wang, L., Dou, Q., Fletcher, P.T., Speidel, S., Li, S. (eds.) MICCAI 2022. LNCS, vol. 13431, pp.113–122. Springer, Cham (2022). https://doi.org/10.1007/978-3-031-16431-6_11
19. Chen, G., Hon,g Y., Huynh, K. M., et al.: Deep learning prediction of diffusion MRI data with microstructure-sensitive loss functions. Med. Image Anal. **85**, 102742 (2023)
20. Gibbons, E.K., Hodgson, K.K., Chaudhari, A.S., et al.: Simultaneous NODDI and GFA parameter map generation from subsampled q-space imaging using deep learning. Magn. Reson. Med. **81**(4), 2399–2411 (2019)

21. Wang, S., Wu, R., Jia, S., et al.: Knowledge-driven deep learning for fast MR imaging: undersampled MR image reconstruction from supervised to un-supervised learning. Magn. Reson. Med. **92**(2), 496–518 (2024)
22. Wang, S., Xiao, T., Liu, Q., Zheng, H.: Deep learning for fast MR imaging: A review for learning reconstruction from incomplete k-space data. Biomed. Signal Process. Control **68**, 102579 (2021)
23. Wang, S., Wu, R., Li, C.: Parcel: physics-based unsupervised contrastive representation learning for multi-coil MR imaging. IEEE/ACM Trans. Comput. Biol. Bioinf. **20**(5), 2659–2670 (2022)
24. Van Essen, D.C., Smith, S.M., Barch, D.M., et al.: The WU-MINN human connectome project: an overview. Neuroimage **80**, 62–79 (2013)
25. Wu, R., Li, C., Zou, J., et al.: Generalizable reconstruction for accelerating MR imaging via federated learning with neural architecture search. IEEE Trans. Med. Imaging (2024)
26. Caruyer, E., Lenglet, C., Sapiro, G., et al.: Design of multishell sampling schemes with uniform coverage in diffusion MRI. Magn. Reson. Med. **69**(6), 1534–1540 (2013)
27. Cheng, J., Shen, D., Yap, PT.: Designing single- and multiple-shell sampling schemes for diffusion MRI using spherical code. In: Golland, P., Hata, N., Barillot, C., Hornegger, J., Howe, R. (eds.) MICCAI 2014. LNCS, vol. 8675, pp. 281–288. Springer, Cham (2023). https://doi.org/10.1007/978-3-319-10443-0_36
28. Cheng, J., Shen, D., Yap, P.T., et al.: Single-and multiple-shell uniform sampling schemes for diffusion MRI using spherical codes. IEEE Trans. Med. Imaging **37**(1), 185–199 (2017)
29. Daducci, A., Canales-Rodríguez, E.J., Zhang, H., et al.: Accelerated microstructure imaging via convex optimization (AMICO) from diffusion MRI data. Neuroimage **105**, 32–44 (2015)
30. Park, J., Jung, W., Choi, E.J., et al.: DIFFnet: diffusion parameter mapping network generalized for input diffusion gradient schemes and b-value. IEEE Trans. Med. Imaging **41**(2), 491–499 (2021)

Introducing QuantConn: Overcoming Challenging Diffusion Acquisitions with Harmonization

Nancy Newlin[1(✉)], Kurt Schilling[2], Serge Koudoro[3], Bramsh Qamar Chandio[4], Praitayini Kanakaraj[1], Daniel Moyer[1], Claire E. Kelly[5,6,7], Sila Genc[8,9], Joseph Yuan-Mou Yang[5,8,9,10], Ye Wu[11], Nagesh Adluru[12], Vishwesh Nath[13], Sudhir Pathak[14], Walter Schneider[14], Anurag Gade[15], William Consagra[16], Yogesh Rathi[16], Tom Hendriks[17], Anna Vilanova[17], Maxime Chamberland[17], Tomasz Pieciak[18,19], Dominika Ciupek[19], Antonio Tristán Vega[18], Santiago Aja-Fernández[18], Maciej Malawski[19], Gani Ouedraogo[20], Julia Machnio[19], Paul M. Thompson[4], Neda Jahanshad[4], Eleftherios Garyfallidis[3], and Bennett Landman[21,22]

[1] Department of Computer Science, Vanderbilt University, Nashville, TN, USA
nancy.r.newlin@vanderbilt.edu
[2] Department of Radiology and Radiological Sciences, Vanderbilt University Medical Center, Nashville, USA
[3] Indiana University Bloomington, Bloomington, IN, USA
[4] Mark and Mary Stevens Neuroimaging and Informatics Institute, Keck School of Medicine of USC, Los Angeles, CA, USA
[5] Developmental Imaging, Murdoch Children' Research Institute, Melbourne, Australia
[6] Victorian Infant Brain Study (VIBeS), Murdoch Children's Research Institute, Melbourne, Australia
[7] Turner Institute for Brain and Mental Health, School of Psychological Sciences, Monash University, Melbourne, Australia
[8] Neuroscience Advanced Clinical Imaging Service (NACIS), Department of Neurosurgery, Royal Children's Hospital, Melbourne, Australia
[9] Neuroscience Research, Murdoch Children's Research Institute, Melbourne, Australia
[10] Department of Paediatrics, University of Melbourne, Melbourne, Australia
[11] School of Computer Science and Technology, Nanjing University of Science and Technology, Nanjing, China
[12] Waisman Center, Department of Radiology, University of Wisconsin, Madison, USA
[13] NVIDIA, Santa Clara, USA
[14] Learning Research and Development Center, University of Pittsburgh, Pittsburgh, USA
[15] Brigham and Women's Hospital, Boston, USA
[16] Brigham and Women's Hospital, Harvard Medical School, Boston, USA
[17] Department of Computer Science and Mathmatics, Eindhoven University of Technology, Eindhoven, The Netherlands
[18] LPI, ETSI Telecomunicación, Universidad de Valladolid, Castilla y León, Spain
[19] Sano Centre for Computational Medicine, 30-054 Kraków, Poland

[20] Aix-Marseille Université, Marseille, France
[21] Vanderbilt University Institute of Imaging Science, Vanderbilt University, Nashville, TN, USA
[22] Department of Electrical and Computer Engineering, Vanderbilt University, Nashville, TN, USA

Abstract. White matter alterations are increasingly implicated in neurological diseases and their progression. Diffusion-weighted magnetic resonance imaging (DW-MRI) has been included in many international-scale studies to identify alterations in white matter microstructure and connectivity. Yet, quantitative investigation of DW-MRI data is hindered by a lack of consistency due to variations in acquisition protocols, sites, and scanners. Specifically, there is a need to harmonize the preprocessing of DW-MRI datasets to ensure that compatible and reproducible quantitative metrics are derived from each site, including (1) bundle-wise microstructure measures, (2) features of white matter fiber bundles, and (3) connectomics measures. In the MICCAI CDMRI 2023 QuantConn challenge, participants are provided raw data from the same individuals taken with two different acquisition protocols on a single 4 tesla scanner in the same scanning session and asked to preprocess the data in order to minimize acquisition differences while retaining biological variation. Here, we outline the testing framework, provide baseline pre-harmonized results, and discuss the learning implications of this challenge.

Keywords: Diffusion MRI · harmonization · macrostructure · microstructure · tractography · connectome · image processing

1 Introduction

Diffusion-weighted magnetic resonance imaging (DW-MRI) is a non-invasive imaging modality that enables in-vivo modeling of white matter microstructure and facilitates structural brain connectivity mapping. DW-MRI has been included in many international-scale studies focused on white matter biomarkers in development, disease, and disorders. Yet, multi-site investigation of DW-MRI data is hindered by a lack of consistency due to variations in acquisition protocols, scanner settings, and manufacturers [17]. Substantial progress has been made with calibration and harmonization to reduce inter-subject variance and improve the interpretability of computed measures. However, the fundamental challenge remains that the clinical application of DW-MRI (as currently implemented) is confounded by inter-scanner and inter-site effects. Here, we introduce the QuantConn challenge - an international community challenge which aims to harmonize differences in diffusion acquisitions.

This challenge builds off the successful SuperMUDI [15] and MUSHAC [13] challenges. In SuperMUDI, participants were tasked with super-resolving data (N=5 volunteers) that have high in-plane resolution but thick (axial) slices in order to match an isotropically acquired dataset. In the MUSHAC challenge, participants were given datasets (N=15 volunteers) that were acquired on two different scanners, with two different acquisitions on each, and tasked with minimizing cross-scanner and cross-protocol differences in voxel-wise indices. Here, we provide 103 pairs of datasets of the same subjects, scanned with very different acquisition protocols - and challenge participants to minimize differences in the data in order to minimize differences in microstructure, tractography bundle analysis, and connectomics measures. The key innovations of the Quant-Conn challenge are (1) we assess bundles and tractography in the context of harmonization for the first time, (2) we assess connectomics in the context of harmonization for the first time, and (3) we have 10x additional subjects over MUSHAC and 100x over SuperMUDI. Additionally, the data that form the basis of this challenge represent a difficult clinical scenario for harmonization and are part of a much larger twins study, which could provide rich context for continuing validation and extension of this challenge's findings.

In the following, we introduce the challenge, describe the data and evaluation framework, and present baseline measures against which challenge submissions will be compared against. We end with a discussion on what we aim to investigate, and hope to learn from this community effort.

1.1 The Challenge

Participants are challenged to bridge acquisition disparities in scan-rescan DW-MRI data - using potentially preprocessing, super-resolution, and/or harmonization techniques. They are provided with a T1-weighted MRI and two DW-MRI scans for a cohort of 78 training and 25 testing patients. The core differences between the two acquisitions are (1) spatial resolution ($1.8 \times 1.8 \times 5$ mm^3 vs. $1.8 \times 1.8 \times 2$ mm^3) and (2) number of directions collected (27 vs. 94). These differences bias and confound three major groups of diffusion analysis: bundle microstructure, bundle macrostructure, and complex network measures (Fig. 1) [11,17,20,21]. A successful challenge algorithm will minimize protocol differences in all three analyses.

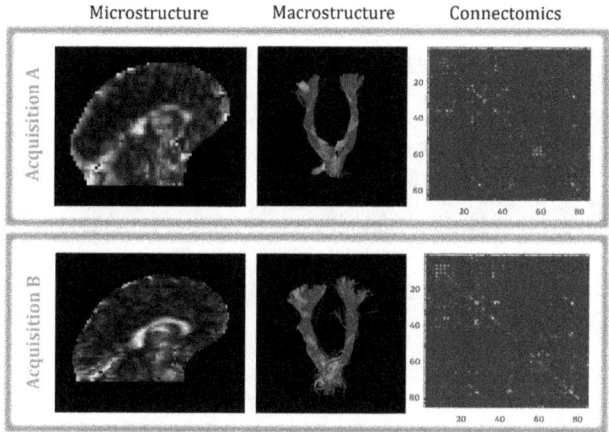

Fig. 1. Acquisition differences affect microstructure measurements (FA map, left), macrostructure reconstructions (corticospinal tract shape, middle), and connectomics (whole brain connectome, right).

2 Data

The data that form the basis of this challenge represent a difficult clinical scenario for harmonization and are a subset of the Queensland Twin Imaging study [19]. The DW dataset consists of 25 matched testing and 78 training subjects scanned twice with two different acquisition protocols (Fig. 2). Testing and Training data used in this challenge are publicly available at shared Box. Each subject has an anatomical 3D whole-brain T1-weighted image. The data subset is comprised of 45% females, ages 25.3 ± 1.8 years. No subjects reported a history of significant head injury, neurological or psychiatric illness, or substance abuse or dependence, and no subjects had a first-degree relative with a psychiatric disorder. All subjects were right-handed as determined using 12 items from Annett's Handedness Questionnaire [2]. Scanning was performed at the QIMR Berghofer Medical Research Institute on a 4 tesla Siemens Bruker Medspec scanner [19].

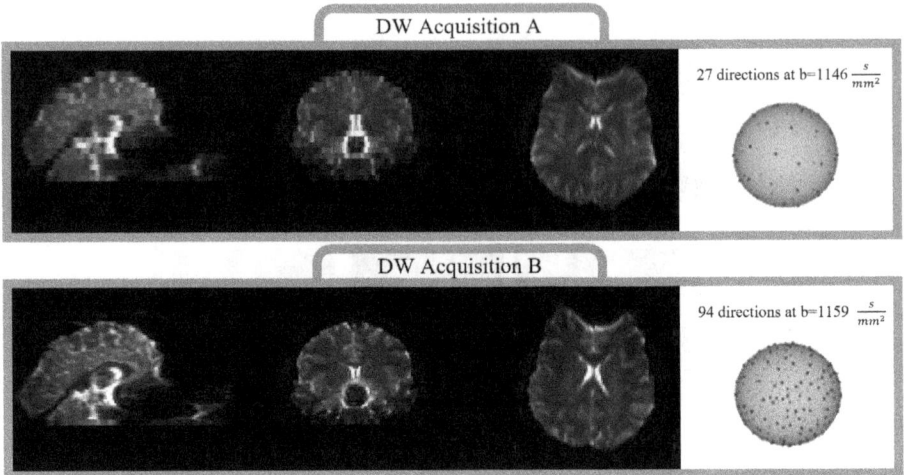

Fig. 2. Each subject was scanned using two protocols with different resolutions (left) and gradient directions (right). Image resolution and quality differences are apparent in visualizations of b0-images (b = 0 s/mm²).

2.1 Acquisition A

Acquisition A DW images were acquired using single-shot echo-planar imaging with a twice-refocused spin echo sequence. Imaging parameters were repetition/echo times of 6090/91.7 ms, field of view of 23 cm, and 128×128 acquisition matrix. Each 3D volume consisted of 21 axial slices 5 mm thick with a 0.5 mm gap and 1.8×1.8 mm² in-plane resolution, total time = 3 min. Thirty images were acquired per subject: three with no diffusion sensitization (b = 0 s/mm²) and 27 DW images (b = 1146 s/mm²) with gradient directions uniformly distributed on the hemisphere.

2.2 Acquisition B

Aquisition B DW images were acquired using single-shot echo planar imaging (EPI) with a twice-refocused spin echo sequence. Imaging parameters were: 23scm FOV, TR/TE 6090/91.7 ms, with a 128×128 acquisition matrix. Each 3D volume consisted of 55 2-mm thick axial slices with no gap and a 1.79×1.79 mm² in-plane resolution with total acquisition time = 14.2 min. 105 images were acquired per subject: 11 with no diffusion sensitization (b=0 s/mm²) and 94 DWI (b = 1159 s/mm²) with gradient directions uniformly distributed on the hemisphere.

3 Testing Framework

Participants will submit 25 paired DW-MRI from the testing cohort. Submission processing includes: diffusion analysis of tensor fitting, orientation distribution function (ODF) reconstruction, whole brain tractography, bundle segmentation, tractometry, bundle shape analysis, connectomics, and finally, complex network analysis. Code is available via a public docker. Testing data are 2.9 GB and evaluation takes 9 h on a 64 GB RAM Macbook Pro.

Assessment will be performed to quantify the similarity of (1) bundle-specific microstructure measures, (2) macrostructural features of each pathway, and (3) complex network measures. Table 1 reports the baseline scores for each category.

Table 1. Baseline scores for un-harmonized data.

Connectivity score	Microstructural score	Shape similarity score	Shape profile score
0.37	0.19	0.85	6.79

3.1 Bundle-Specific Microstructure

Microstructure measures of fractional anisotropy (FA), mean diffusivity (MD), geodesic anisotropy (GA), radial diffusivity (RD), and axial diffusivity (AD) are calculated for each of six pathways as the average DTI measure within each bundle: left/right Arcuate Fasciculus (AF), left/right Optic Radiations (OR), and left/right Corticospinal tract bundles (CST)

Ideally, DTI in these ROIs would be the same across scans for a single patient. However, Fig. 1 shows bundles are not consistent across acquisitions A and B. We use Intra-class Correlation Coefficient (ICC) to quantify reproducibility across acquisitions and bundles. ICC quantifies how well measurements of the same group resemble each other. The QIBA Technical Performance Working Group suggests using ICC to evaluate repeatability. The ICC is interpreted as follows: below 0.5 as "poor", between 0.50 and 0.75 as "moderate", between 0.75 and 0.90 as "good", and above 0.90 as "excellent" [10]. We evaluate the repeatability of diffusion measures for scan-rescan of the same subject, for all 25 subjects in the testing cohort. Baseline results for ICC per bundle, before any preprocessing or harmonization solution has been applied, is reported in Fig. 3. We compute a single microstructural score that is the average ICC across bundles. Successful submissions will increase ICC and have a higher microstructural score than that reported in Table 1.

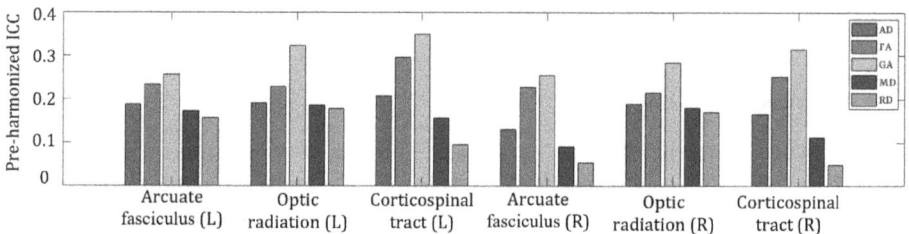

Fig. 3. ICC, computed across 25 A/B pairs, of bundle microstructural measures AD, FA, GA, MD, RD for 6 bundles. Successful submissions will improve consistency of microstructural measures and have higher ICC values.

3.2 Bundle-Specific Macrostructure

Using the same six bundles above, we use the BUndle ANalytics (BUAN) framework to compare the reconstructions created from each acquisitions' DW-MRI [5],[?].

BUAN provides bundle shape similarity scores to assess structural consistency among bundles. Bundle shape similarity is a streamline-based comparison approach that provides a single score between 0 (no match) and 1 (perfect match) between two bundles. Furthermore, we created bundle shape profiles using BundleWarp registration [4]. The bundle shape profile describes shape differences in bundles across the two acquisitions in terms of deformation magnitude generated by BundleWarp registration to match two bundles completely. Lower deformation magnitude implies fewer shape differences in two bundles, and higher deformation magnitude means a high number of deformations and/or transformations are required to align two bundles perfectly, hence having higher shape differences.

Bundle shapes should be consistent across scans of the same patient. Thus, we expect a successful submission will produce DW-MRI that has low bundle shape profiles and high shape similarity scores across acquisitions. The bundle shape profile and shape similarity scores are reported for 25 Test patients in Fig. 4. The final bundle-based score is the average across all 25 patients.

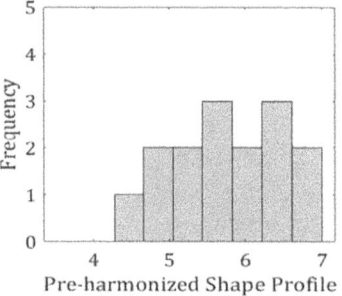
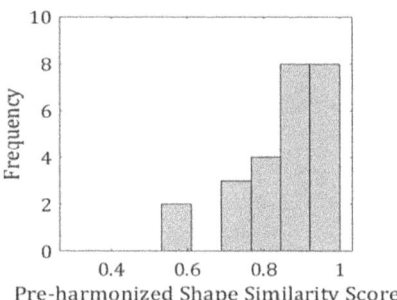

Fig. 4. BundleWarp macrostructure comparison measures, bundle shape profiles (left) and bundle shape similarity score (right), computed across acquisitions for all 25 test-set patients. Successful submissions will improve DW-MRI quality such that bundle shape similarities are higher and bundle shape profiles are lower.

3.3 Connectome Measures

The last class of metrics that we derive are complex network measures computed on the connectome. To construct a connectome, we first reconstruct ODFs using the Constant Solid Angle (CSA) algorithm at a maximum spherical harmonic order of 6 [1]. Using the ODF and a white matter mask ($FA > 0.1$), we perform whole-brain tractography using the EuDX [7] algorithm. The connectome is the result of segmenting and mapping the whole brain tractogram into brain regions (nodes) and the streamlines connecting them (edges). For this challenge, tractograms are segmented based on the Desikan-Killiany atlas from FreeSurfer (version 7.2.0) with 87 cortical and subcortical regions [6].

Connectome measures are intended to characterize brain networks in a computationally simple yet meaningful manner [16]. In this challenge, we consider modularity, assortativity, average participation coefficient, density, average node strength, and average clustering coefficient. Modularity is the degree to which the network can be subdivided into clearly delineated subgroups and summarizes the community structure. Assortativity is a correlation coefficient between nodes of opposite ends of a link and reflects the network's resilience. The participation coefficient is a measurement of the diversity of intermodular connections of nodes. Density is the fraction of present connections to total possible connections. Node strength is the sum of weights for edges connected to a given node. The clustering coefficient is the geometric mean associated with a node. We use the scilpy (version 1.5.0) and bctpy (version 0.6.1) implementations of the Brain Connectivity Toolbox measure definitions [16].

A successful submission would produce DW-MRI with consistent complex network measures across acquisitions of the same patient. Figure 5 is a histogram report highlighting the significant baseline differences between acquisitions. We assess reproducibility across complex network measures with ICC. This ICC is the final connectivity score (Table 1).

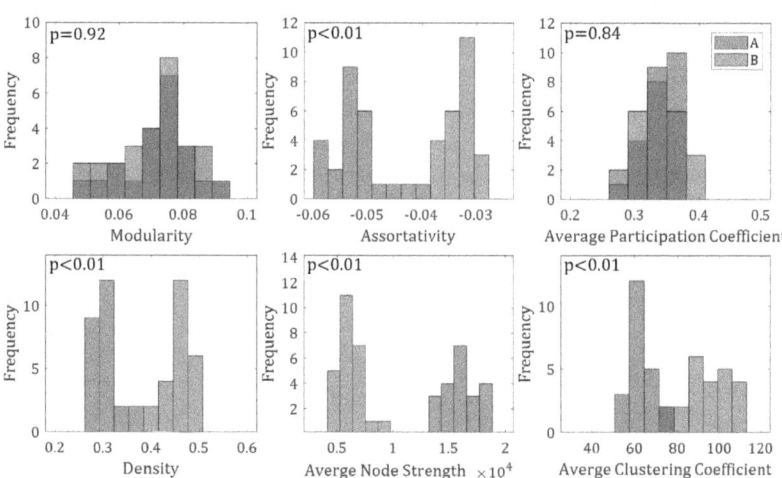

Fig. 5. Histogram distributions for connectome measures derived from pre-harmonized A and B DWI acquisitions. t-test p-values for comparing means across the two acquisitions are reported in the top left corner.

4 Broader Contributions

Data are increasingly coming from multi-center, multi-site, and multi-acquisition sources, and it is important to utilize these datasets to detect changes in development [3,9,18], Alzheimer's disease [12,14], and autism [8]. For this reason, harmonization is particularly important in DW-MRI so that bundles and connectomes are comparable, and we can study features from micro-scale to macro-scale. These features act as potential biomarkers for neurological diseases, and subtle, important changes can be overshadowed by un-harmonized biases.

This challenge gives us insight into the efficacy of diffusion harmonization methods that aim to make one site acquisition like another. Current examples include statistical and image-based methods, which have proven useful so far, but methods are continuously being developed and improved. This study is a benchmark that lets us define method successes and limitations. A successful submission in this challenge is one that overcomes the dramatic differences due to resolution and gradient directions. While this challenge tasks image-based methods of diffusion harmonization, i.e. making the signal from scanners similar, we do not specifically investigate statistical approaches which have become common in literature (ComBat, CovBat, NeuroCombat). As as extension of this challenge and all submissions, we plan to investigate the fidelity and robustness of statistical feature based harmonization methods. For example, we are not aware of these approaches being applied to features of bundles.

Beyond simple harmonization, this challenge also acts as a super-resolution task. The task of super resolution is simply to perform image processing that results in an image with information captured at a higher resolution than it was acquired at. This method is important in tractography to make data isotropic, and is more critical as we try to investigate smaller structures in the brain. A successful submission in this challenge will necessitate super-resolving the low resolution dataset to generate a higher, isotropic resolution dataset that closely matches signal actually acquired at a higher spatial resolution.

Finally, preprocessing is necessary to remove effects and biases and improve SNR. For example, preprocessing mitigates motion, susceptibility and eddy current distortions - a necessary step for robust, reproducible measurements. Because these data have different acquisitions and, therefore, different artifacts and distortions, this is a study on preprocessing techniques that minimize differences due to imaging artifacts.

Fig. 6. Twenty teams from across the world have registered to participate in the Quant-Conn challenge. This map visualization was made with mapchart.com.

We expect participants (Fig. 6) to submit a range of methods such as explicit signal harmonization, resolution adjustments, and machine learning solutions.

Acknowledgements. The Queensland Twin Imaging study data collection was funded by National Institute of Child Health and Human Development (R01 HD050735), and National Health and Medical Research Council (496682, 1009064).

This work was also supported in part by NIH grants R01EB017230, NIDDK K01EB032898, R01NS123378, P50HD105353, RF1AG057892 (FiberNET project grant), R01 MH134004 and R01MH119222 (PIs: Rathi, O'Donnell).

Tomasz Pieciak acknowledges the Polish National Agency for Academic Exchange for grant PPN/BEK/2019/1/00421 under the Bekker programme and the Ministry of Science and Higher Education (Poland) under the scholarship for outstanding young scientists (692/STYP/13/2018). Tomasz Pieciak, Antonio Tristán Vega and Santiago Aja-Fernández were supported by research grants PID2021-124407NBI00, funded by MCIN/AEI/10.13039/501100011033/FEDER, UE, and TED2021-130758B-I00, funded by MCIN/AEI/10.13039/501100011033 and the European Union "NextGeneration EU/PRTR". Dominika Ciupek, Maciej Malawski and Julia Machnio were supported by the European Union's Horizon 2020 research and innovation program under grant agreement Sano No 857533 and the International Research Agendas program of the Foundation for Polish Science No MAB PLUS/2019/13.

JYMY and SG received positional funding from the Royal Children's Hospital Foundation (RCHF 2022-1402). CEK was supported by an Australian Government Research Training Program (RTP) Scholarship, Monash University (Monash Graduate Excellence Scholarship), and the Australian National Health and Medical Research Council (NHMRC; Centre of Research Excellence in Newborn Medicine 1153176). JYMY, SG, and CEK acknowledge the support of the Royal Children's Hospital, Murdoch Children's Research Institute, The University of Melbourne Department of Paediatrics, and the Victorian Government's Operational Infrastructure Support Program. Ye Wu was supported by the National Natural Science Foundation of China (No. 62201265). The content is solely the responsibility of the authors and does not necessarily represent the official views of the NIH.

References

1. Aganj, I., Lenglet, C., Sapiro, G.: Odf reconstruction in q-ball imaging with solid angle consideration. In: Proceedings/IEEE International Symposium on Biomedical Imaging: From Nano to Macro. IEEE International Symposium on Biomedical Imaging 2009, vol. 1398 (2009). https://doi.org/10.1109/ISBI.2009.5193327
2. Annett, M.: A classification of hand preference by association analysis. Brit. J. Psychol. **61**, 303–321 (1970). https://doi.org/10.1111/J.2044-8295.1970.TB01248.X
3. Cetin-Karayumak, S., Zhang, F., O'Donnell, L.J., Rathi, Y.: Harmonization of multi-site diffusion magnetic resonance imaging data from the adolescent brain cognitive development study. Biol. Psychiat. **91**, S84 (2022). https://doi.org/10.1016/j.biopsych.2022.02.227
4. Chandio, B.Q., Olivetti, E., Romero, D., Harezlak, J., Garyfallidis, E.: Bundlewarp, streamline-based nonlinear registration of white matter tracts. In: bioRxiv, pp. 2023–01 (2023)

5. Chandio, B.Q., et al.: Bundle analytics, a computational framework for investigating the shapes and profiles of brain pathways across populations (2020). https://doi.org/10.1038/s41598-020-74054-4
6. Fischl, F.B.: Freesurfer (2012). https://doi.org/10.1016/j.neuroimage.2012.01.021
7. Garyfallidis, E.: Towards an accurate brain tractography [phd thesis]. University of Cambridge, United Kingdom (2012)
8. Heinsfeld, A.S., Franco, A.R., Craddock, R.C., Buchweitz, A., Meneguzzi, F.: Identification of autism spectrum disorder using deep learning and the abide dataset. NeuroImage: Clin. **17**, 16 (2018). https://doi.org/10.1016/J.NICL.2017.08.017
9. Jernigan, T.L., et al.: The pediatric imaging, neurocognition, and genetics (ping) data repository. NeuroImage **124**, 1149 (2016). https://doi.org/10.1016/J.NEUROIMAGE.2015.04.057
10. Koo, T.K., Li, M.Y.: A guideline of selecting and reporting intraclass correlation coefficients for reliability research. J. Chiropractic Med. **15**, 155–163 (2016). https://doi.org/10.1016/J.JCM.2016.02.012
11. Magnotta, V.A., et al.: Multicenter reliability of diffusion tensor imaging. Brain Connect. **2**, 345 (2012). https://doi.org/10.1089/BRAIN.2012.0112
12. Marcus, D.S., Wang, T.H., Parker, J., Csernansky, J.G., Morris, J.C., Buckner, R.L.: Open access series of imaging studies (oasis): cross-sectional MRI data in young, middle aged, nondemented, and demented older adults. J. Cogn. Neurosci. **19**, 1498–1507 (2007). https://doi.org/10.1162/JOCN.2007.19.9.1498
13. Ning, L., et al.: Muti-shell diffusion mri harmonisation and enhancement challenge (mushac): Progress and results. In: Mathematics and Visualization, pp. 217–224 (2019). https://doi.org/10.1007/978-3-030-05831-9_18/COVER
14. Petersen, R.C., et al.: Alzheimer's disease neuroimaging initiative (adni): clinical characterization. Neurology **74**, 201 (2010). https://doi.org/10.1212/WNL.0B013E3181CB3E25
15. Pizzolato, M., Palombo, M., Hutter, J., Nash, V., Zhang, F., Gyori, N.: Super-resolution of multi dimensional diffusion mri data (2020). https://doi.org/10.5281/ZENODO.3718990
16. Rubinov, M., Sporns, O.: Complex network measures of brain connectivity: uses and interpretations. NeuroImage **52**, 1059–1069 (2010). https://doi.org/10.1016/J.NEUROIMAGE.2009.10.003
17. Schilling, K.G., et al.: Fiber tractography bundle segmentation depends on scanner effects, vendor effects, acquisition resolution, diffusion sampling scheme, diffusion sensitization, and bundle segmentation workflow. NeuroImage **242**, 118451 (2021). https://doi.org/10.1016/J.NEUROIMAGE.2021.118451
18. Somerville, L.H., et al.: The lifespan human connectome project in development: a large-scale study of brain connectivity development in 5–21 year olds. NeuroImage **183**, 456 (2018). https://doi.org/10.1016/J.NEUROIMAGE.2018.08.050
19. Strike, L.T., et al.: queensland twin imaging (qtim) (2023). https://doi.org/10.18112/openneuro.ds004169.v1.0.7
20. Tax, C.M., et al.: Cross-scanner and cross-protocol diffusion mri data harmonisation: a benchmark database and evaluation of algorithms. NeuroImage **195**, 285–299 (2019). https://doi.org/10.1016/j.neuroimage.2019.01.077
21. Vollmar, C., et al.: Identical, but not the same: intra-site and inter-site reproducibility of fractional anisotropy measures on two 3.0 t scanners. NeuroImage **51**, 1384–1394 (2010). https://doi.org/10.1016/J.NEUROIMAGE.2010.03.046

Learning Low-Rank Tensor Approximation for GPU-Based Tractography

Johannes Gruen[1] and Thomas Schultz[1,2]

[1] B-IT and Institute for Computer Science, University of Bonn, Bonn, Germany
schultz@cs.uni-bonn.de
[2] Lamarr Institute for Machine Learning and Artificial Intelligence, Bonn, Germany

Abstract. Fast algorithms for diffusion MRI tractography are required due to the increasing amounts of diffusion MRI data, and the increasing popularity of whole-brain tractography. Representing fiber orientation density functions (fODFs) as higher-order tensors and extracting main fiber directions from them via low-rank tensor approximation is a state-of-the-art variant of streamline tractography, but involves a computationally costly nonlinear optimization in each integration step. In this work, we demonstrate that unsupervised training of a neural network to map fODF coefficients to the corresponding fiber contributions directly is not only faster, but also achieves lower approximation residuals. This is due to the fact that training the network amounts to a joint optimization of all fiber contributions, while traditional algorithms follow an alternating optimization strategy. However, we observe that the traditional approach implicitly favors sparse solutions, and that a corresponding explicit regularization is required to obtain useful results with a joint optimization strategy. Building on those insights, we create the first GPU-based implementation of low-rank tractography, which achieves a speedup by a factor of 68, compared to traditional tractography on a single CPU core, while at the same time improving the median dice.

Keywords: Tractography · Low-rank Tensor Approximation · Deep Neural Networks

1 Introduction

Diffusion MRI tractography [9] is a unique in-vivo method used to reconstruct white matter fiber tracts and plays a crucial role in surgery planning [2] and scientific studies [3,4]. Typically, Constrained Spherical Deconvolution (CSD) is

---

Funded by the Federal Ministry of Education and Research within the project "BNTrAinee" (funding code 16DHBK1022). Data were provided by the Human Connectome Project, WU-Minn Consortium (Principal Investigators: David Van Essen and Kamil Ugurbil; 1U54MH091657) funded by the 16 NIH Institutes and Centers that support the NIH Blueprint for Neuroscience Research; and by the McDonnell Center for Systems Neuroscience at Washington University.

applied to estimate the fiber Orientation Distribution Function (fODF) at voxel level [5]. Using local integration schemes, a fiber bundle representation can be reconstructed. Representing the fODFs as higher-order tensors and estimating a low-rank approximation has been shown to yield higher angular accuracy, leading to state-of-the-art tractography results [1,7].

However, the proposed low-rank approximation method is time consuming, which becomes a major drawback with increasing sizes of datasets. In this work, we propose a learned low-rank approximation method with two main advantages. Firstly, it is faster, as the complex optimization process is replaced by a forward pass through a neural network and we can effectively parallelize over all seeds on a GPU. Secondly, unlike the traditional approach where low-rank optimization is computed iteratively for each peak, our method estimates all fiber directions simultaneously, resulting in higher angular accuracy. To achieve this, we introduce a new regularization in Sect. 3 to address the ill-conditioning of the low-rank approximation.

The results presented in Sect. 4 indicate that the learned low-rank approximation significantly reduces residual errors compared to the classical low-rank approximation, but require only a fraction of the time. Finally, our new regularizer allows us to effectively tune the amount of fanning in reconstructions. Depending on its weight, tractography results yield a higher Dice compared to those obtained with traditional low-rank approximation.

2 Related Work and Background

The general strategy of training deep neural networks to replace expensive iterative optimization has been used previously in computational diffusion MRI. This included fitting models of diffusion or tissue microstructure [6] and fiber Orientation Distribution Function (fODF) estimation [10]. The use of GPUs to accelerate tractography goes back even further [11], and a GPU framework that supports modeling as well as tractography has been described [8]. Our work is the first to apply this idea to low-rank approximation based tractography [7,13]. It highlights an ambiguity that arises in that context, and proposes a novel regularizer to resolve it.

Low-rank approximation based tractography can be understood as a variant of Constrained Spherical Deconvolution (CSD) [14], which computes an fODF, capturing the fraction of fibers in any direction, and estimates the principal fiber directions from its local maxima. In our work, the fODF is estimated as a symmetric fourth-order tensor $\mathcal{T}$, which involves a modified deconvolution kernel, and allows one to enforce a strict non-negativity constraint, as described by Ankele et al. [1]. From a symmetric fourth order Tensor $\mathcal{T}$, r fiber directions are estimated via a rank-r approximation

$$\mathcal{T}^{(r)} = \sum_{i=1}^{r} \alpha_i \mathbf{v}_i^{\otimes 4}, \tag{1}$$

where the scalar $\alpha_i \in \mathbb{R}_+$ denotes the volume fraction of the ith fiber, $\mathbf{v}_i \in \mathbb{S}^2$ its direction, and the superscript $\otimes 4$ indicates a 4-fold symmetric outer product, which turns the vector into a fourth-order tensor [13].

Prior work has shown benefits of this approach compared to classical CSD. In particular, it is able to separate crossing fibers even if they are not distinct maxima. Consequently, fourth-order tensor approximation yields a higher angular resolution than eighth-order fODFs with peak extraction, while the reduced order leads to better numerical conditioning of the fODF estimation [1]. In this work, we develop an efficient strategy for computing the low-rank approximation.

3 Methods and Material

3.1 Learning the Low-Rank Approximation

Low-rank approximation finds volume fractions $\alpha_i \in \mathbb{R}+$ and fiber directions $\mathbf{v}_i \in \mathbb{S}^2$, $i \in \{1, \ldots, r\}$, so that the resulting rank-r fODF tensor $\mathcal{T}^{(r)}$ according to Equation (1) best approximates the fODF tensor $\mathcal{T}$ that has been estimated from the diffusion MRI data with respect to the Frobenius norm, $\|\mathcal{T} - \mathcal{T}^{(r)}\|_F$.

This approximation is traditionally solved with iterative gradient-based numerical optimization, which is time consuming: Recent work reported 43 s to compute 1000 streamlines based on it [7]. Therefore, we instead train a multilayer perceptron (MLP) to map 15 input channels, representing the unique coefficients of $\mathcal{T}$, to 9 output values, representing three scaled directions that correspond to the terms of a rank-three approximation $\mathcal{T}^{(3)}$. Three is commonly assumed to be an upper bound for the number of distinct fiber compartments that can be reliably estimated within a single voxel [12].

Since training the MLP on simulated fODFs for which the true directions and volume fractions are known might not generalize to real fODFs, we train it in an unsupervised manner, by directly optimizing the approximation error $\|\mathcal{T} - \mathcal{T}^{(3)}\|_F$ on real fODFs. This corresponds to a simultaneous optimization of all three fiber compartments from random initializations, which provides additional flexibility compared to the optimization algorithm that has been used previously, which simplified and sped up the process with an alternating approach that refines the compartments one by one, fixing the remaining ones [13].

Moreover, the optimization now corresponds to training a neural network that, once suitable weights for all plausible fODFs have been found, can be applied for performing low-rank approximation with a simple forward pass. Therefore, we can afford more stringent termination criteria for training than previously used for per-voxel optimization. For these two reasons, we can expect the MLP approach to achieve lower approximation errors than the previously used optimization.

However, this approach also reveals concerns regarding a potential lack of uniqueness in the optimization that have been implicitly avoided previously. They are most easily illustrated if we assume that the rank of the true fODF is lower than the approximation rank. For example, assume an fODF with a single fiber which has volume fraction one. Since this fODF has rank one, multiple rank-3 representations with zero approximation residuals can be found, with parallel directions and arbitrary non-negative volume fractions that add up to one. When optimizing in an alternating way, this ambiguity is resolved automatically: Optimizing the main fiber in isolation will assign it a volume fraction of one, and the second and third fiber remain at their initial values of zero. When optimizing simultaneously from a random initialization, we have to account for this case explicitly.

Due to fiber curvature, spread, and measurement noise, fODFs estimated from real data are unlikely to have rank lower than three, even in so-called single-fiber voxels. However, we frequently observed that increasing the accuracy of the low-rank approximation in such voxels led to two or three near-parallel detections, which correspond to the same white matter tract, and should be treated as one for the purpose of tractography. We propose to resolve these problems by introducing a new regularization term that penalizes this behavior. The optimization problem that serves as a loss for training of our MLP thus becomes

$$\min_{\alpha_i, \mathbf{v}_i} \|\mathcal{T} - \sum_{i=1}^{r} \alpha_i \mathbf{v}_i^{\otimes 4}\|_F + \nu \sum_{i=1}^{r} \frac{1}{1 + \exp\left(10 \log\left(\frac{1}{99}\right) \alpha_i\right)}, \quad (2)$$

where ν is a newly introduced regularization parameter. In our experiments, we evaluated $\nu = 0.02$ and $\nu = 0.05$, and found that ν can be used as a tuning parameter, with larger values reducing anatomically implausible streamlines, but also the amount of reconstructed fanning. Given that deconvolution leads to a similar scaling of fODFs across datasets, similar values of ν should also work on other datasets. The regularizer itself is a sigmoid function that is tuned such that it is almost saturated at 0.1. This draws volume fractions $\alpha_i < 0.1$ towards zero, which resolves the problem of redundant reconstructions: Compared to a single fiber compartment, multiple near-parallel fibers result in a small reduction in approximation error, which is more than compensated by the penalty that the regularizer imposes on each non-vanishing compartment. Consequently, the desired single fiber solution now leads to the lowest loss. As a side effect, this avoids reconstructing fibers with small volume fractions, which are often noise-related artifacts, and not used for tractography. A widely used algorithm for constrained deconvolution similarly penalizes volume fractions below 0.1 [14].

Our experiments use a simple MLP with 5 hidden layers with 400 neurons each. Between each layer, a GeLU activation function is used. An AdamW optimizer is employed with a learning rate of 0.001 and a learning rate scheduler which reduces the learning rate by 0.7 if the loss does not improve over 20 epochs. The training is stopped if the validation loss does not improve for 100 epochs.

We use 10 subjects from the Human Connectome Project (HCP) for training and 2 subjects for validation [15]. We augment the input fODFs by applying random flipping along the x, y, and z axes, and by adding random noise. To illustrate that the MLP does not have to be overfitted to individual subjects, the training and validation subjects are distinct from those used for tractography.

3.2 Probabilistic Streamline-Based Tractography

Given a seed point, we initialize the tractography by selecting the direction corresponding to the largest volume fraction and perform Euler integration at each step with a step size of 0.7. At each step, we estimate the low-rank approximation as proposed in Subsect. 3.1. From the three estimated directions, the next direction is randomly drawn. Each direction $\hat{\mathbf{w}}_i$ with length $\hat{\alpha}_i$ is assigned a probability:

$$p_{\alpha,\mathbf{w}}(\hat{\alpha}_i, \hat{\mathbf{w}}_i) := \frac{\mathbf{1}_{\{\Theta_i < \frac{1}{3}\pi\}} \exp\left(-\frac{\|\alpha - \hat{\alpha}_i\|}{2}\right) \cos\left(\left(\frac{3}{\sqrt{2\pi}}\Theta_i\right)^2\right)^6}{\sum_j \mathbf{1}_{\{\Theta_j < \frac{1}{3}\pi\}} \exp\left(-\frac{\|\alpha - \hat{\alpha}_j\|}{2}\right) \cos\left(\left(\frac{3}{\sqrt{2\pi}}\Theta_j\right)^2\right)^6}, \quad (3)$$

based on the previous tracking direction $\mathbf{w}$ with length α, where $\Theta_j = \arccos(\langle \hat{\mathbf{w}}_j, \mathbf{w} \rangle)$ is the angle between the previous and possible directions, and the indicator function $\mathbf{1}$ restricts the choice to angles below 60°. This assigns the highest probabilities to directions most similar to the previous one. Up to an angle of 30°, directions are weighted almost equally, unless one of them matches better in terms of its volume fraction.

This process is repeated until a stopping criterion is reached. We stop the integration if no valid direction is found within 60° or if the partial white matter density estimated from the T1 image drops below 0.3.

3.3 Implementation Details

As computational speed with high accuracy becomes paramount due to the increasing size of datasets, we have decided to implement a GPU-based tractography method. This approach is highly suitable for the iterative integration task, which can be parallelized over all seed points. Given that the computation of streamlines is memory-intensive and GPU memory is commonly smaller than system memory, we opted to keep only the necessary information for computing the next step in the GPU memory, namely the current position and the last direction. Previous points are continuously returned to the system memory. The code is openly available as part of the bonndit package: https://github.com/MedVisBonn/bonndit.

4 Results

The goal of this work is to show that the proposed optimization is more accurate and faster than the previously used iterative optimization.

We show the ability of the MLP framework to solve the low-rank approximation problem by overfitting the MLP on one HCP subject and evaluating how often the learned low-rank approximation yields better results, in terms of

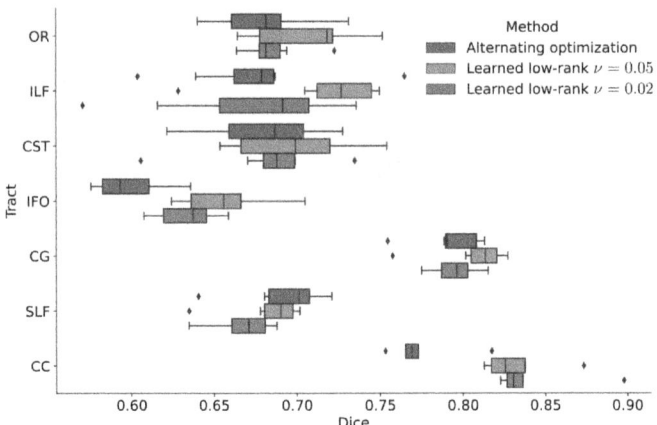

Fig. 1. Comparison of Dice scores between the alternating low-rank optimization and the learned approximation for $\nu = 0.02$ and $\nu = 0.05$. In all tracts except for the SLF, the learned approximation with $\nu = 0.05$ leads to higher median Dice scores than the traditional approach.

optimization residual, than the alternating optimization. It turns out that the learned low-rank approximation yields superior results in 97% of all white matter voxels of that subject for $\nu = 0.05$ and 100% for $\nu = 0.02$. To evaluate the network's generalization capability, we used the trained network and evaluated it on 12 unseen HCP subjects. The learned low-rank approximation leads to smaller residuals in 92% of all white matter voxels for $\nu = 0.05$ and 96% for $\nu = 0.02$. Residuals increase with larger ν, as the regularization favors sparser solutions over fitting the fODFs exactly. Despite this, we conclude that the learned low-rank approximation is more accurate than the classical low-rank approximation for both settings of ν.

To evaluate the capability of the learned low-rank approximation towards tractography, we select 12 HCP subjects with reference reconstructions published as part of TractSeg [17]. The proposed learned low-rank approximations are compared to the alternating low-rank approximation. Therefore, we also apply a bilateral filter to the input fODFs, as suggested in [7]. We seed in a single slice of each tract and perform tractography, as described in Sect. 3.2. To initially start in the right direction, we initialize the tractography with reference directions from TractSeg. The results are filtered using the descriptions from Wakana et al. [16] and evaluated using the Dice similarity coefficient.

As depicted in Fig. 1, the learned low-rank approximation with $\nu = 0.05$ yields the highest median Dice in most cases, followed by the weaker regularized version and the classical low-rank approximation. Exceptions are the CC tract, where the weaker regularized version leads to a higher Dice, and the SLF tract, where the classical method shows the highest Dice. Visual comparison (Fig. 2) of the right Cingulum tract shows that both methods lead to plausible results, and illustrates that the ν parameter is an effective way to control the amount

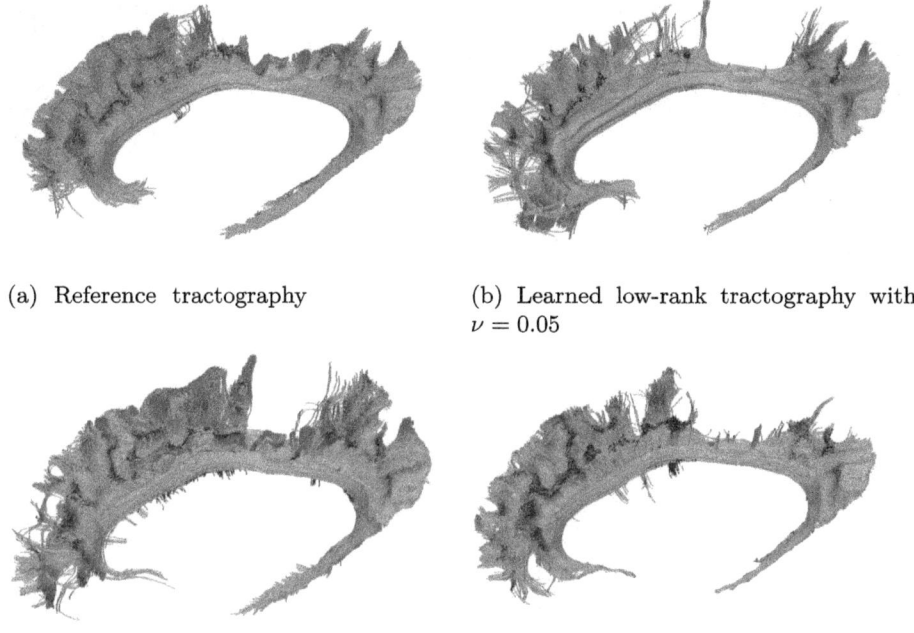

(a) Reference tractography

(b) Learned low-rank tractography with $\nu = 0.05$

(c) Learned low-rank tractography with $\nu = 0.02$

(d) Alternating optimization tractography

Fig. 2. Reconstruction of the right Cingulum in an HCP subject. Both methods reconstruct most of the bundle. The ν parameter allows one to control the amount of fanning in the reconstruction. A small $\nu = 0.02$ leads to strong fanning, comparable to the alternating optimization approach, while the learned low-rank approximation with $\nu = 0.05$ yields more aligned fibers.

of fanning. The weaker regularization leads to more fanning, comparable to the classical low-rank approximation, while the stronger regularization yields more aligned results. This impression gets supported by the number of reconstructed streamlines: With the same number of seeds, 46,000 of the streamlines reconstructed with $\nu = 0.05$ are left after filtering out anatomically implausible ones, but only 38,000 of those tracked with $\nu = 0.02$, indicating a stronger exploration that leads to a more complete reconstruction, but also to more false positives. From the optimization based tractography, 36,000 streamlines are kept.

To show the benefit of the new regularization parameter, we compare the resulting multi-vector fields for $\nu = 0$, $\nu = 0.02$, and $\nu = 0.05$ in Fig. 3. It is visible that the single fiber directions are split into multiple peaks, while the regularized approaches lead to more consistent results for both ν values.

Finally, we evaluate the computation times of the learned low-rank approximation compared to the classical low-rank approximation. Figure 4 shows results for $\nu = 0.05$ and various amounts of seeds. For large numbers of seeds the learned

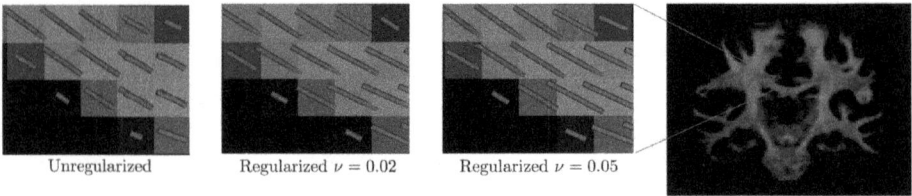

Fig. 3. Comparison of the learned low-rank approximation for $\nu = 0$, $\nu = 0.02$, and $\nu = 0.05$. While the unregularized version leads to splitting of small angles, the regularized versions lead to consistent fiber directions.

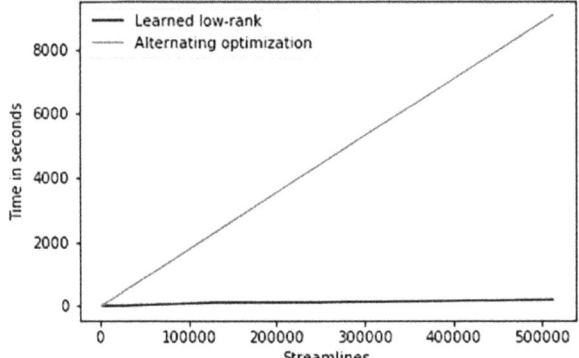

Fig. 4. Comparison of time consumption for tractography. Seeds are placed randomly within the white matter. The time consumption of the optimization based low-rank approximation increases linearly with the seeds. The learning based tractography leads to faster tractography results compared to optimization based tractography.

low-rank tractography was 68 times faster than the optimization based tractography. This evaluation was conducted on a system with a 14-core CPU running at 3.3 GHz and a 2080 Ti GPU. The optimization based tractography runs on a single core and the learned low-rank tractography on the GPU.

5 Conclusion

We developed a GPU-based tractography algorithm that takes only 1.4% of the computation time compared to a CPU-based implementation that computes low-rank approximations of fODF tensors with alternating optimization. A central building block of our algorithm is a learned low-rank tensor approximation, which replaces the time consuming optimization in each streamline integration step with a single forward pass through a lightweight MLP. To achieve this, we introduced a novel regularization term. Our results demonstrate that the trained model generalizes well and produces smaller residual errors than the classical low-rank approximation. For tractography, this leads to a higher Dice score in 6 out of 7 tracts than the alternating low-rank approach. Further, we have shown that tuning ν controls the amount of fanning.

References

1. Ankele, M., Lim, L.-H., Groeschel, S., Schultz, T.: Versatile, robust, and efficient tractography with constrained higher-order tensor fODFs. Int. J. Comput. Assist. Radiol. Surg. **12**(8), 1257–1270 (2017). https://doi.org/10.1007/s11548-017-1593-6
2. Chen, Z., et al.: Corticospinal tract modeling for neurosurgical planning by tracking through regions of peritumoral edema and crossing fibers using two-tensor unscented Kalman filter tractography. Int. J. Comput. Assist. Radiol. Surg. **11**(8), 1475–1486 (2016). https://doi.org/10.1007/s11548-015-1344-5
3. Cheng, G., Salehian, H., Forder, J.R., Vemuri, B.C.: Tractography from HARDI using an intrinsic unscented kalman filter. IEEE Trans. Medical Imaging **34**(1), 298–305 (2015). https://doi.org/10.1109/TMI.2014.2355138
4. Dalamagkas, K., et al.: Individual variations of the human corticospinal tract and its hand-related motor fibers using diffusion MRI tractography. Brain Imaging Behav. **14**(3), 696–714 (2019). https://doi.org/10.1007/s11682-018-0006-y
5. Dell'Acqua, F., Tournier, J.D.: Modelling white matter with spherical deconvolution: How and why? NMR Biomed. **32**(4) (2018). https://doi.org/10.1002/nbm.3945
6. Golkov, V., et al.: q-space deep learning: twelve-fold shorter and model-free diffusion mri scans. IEEE Trans. Med. Imaging **35**(5) (2016). https://doi.org/10.1109/TMI.2016.2551324
7. Grün, J., Gröschel, S., Schultz, T.: Spatially regularized low-rank tensor approximation for accurate and fast tractography. NeuroImage **271** (2023). https://doi.org/10.1016/j.neuroimage.2023.120004
8. Hernandez-Fernandez, M., Reguly, I., Jbabdi, S., Giles, M., Smith, S., Sotiropoulos, S.N.: Using GPUs to accelerate computational diffusion MRI: from microstructure estimation to tractography and connectomes. Neuroimage **188**, 598–615 (2019). https://doi.org/10.1016/j.neuroimage.2018.12.015
9. Jeurissen, B., Descoteaux, M., Mori, S., Leemans, A.: Diffusion MRI fiber tractography of the brain. NMR Biomed. **32**(4), e3785 (2019). https://doi.org/10.1002/nbm.3785
10. Karimi, D., Vasung, L., Jaimes, C., Machado-Rivas, F., Warfield, S.K., Gholipour, A.: Learning to estimate the fiber orientation distribution function from diffusion-weighted MRI. Neuroimage **239**, 118316 (2021). https://doi.org/10.1016/j.neuroimage.2021.118316
11. McGraw, T., Nadar, M.S.: Stochastic DT-MRI connectivity mapping on the GPU. IEEE Trans. Visualizat. Comput. Graph. **13**(6) (2007). https://doi.org/10.1109/TVCG.2007.70597
12. Schultz, T.: Learning a reliable estimate of the number of fiber directions in diffusion MRI. In: Ayache, N., Delingette, H., Golland, P., Mori, K. (eds.) MICCAI 2012. LNCS, vol. 7512, pp. 493–500. Springer, Heidelberg (2012). https://doi.org/10.1007/978-3-642-33454-2_61
13. Schultz, T., Seidel, H.P.: Estimating crossing fibers: a tensor decomposition approach. IEEE Trans. Visual Comput. Graph. **14**(6), 1635–1642 (2008). https://doi.org/10.1109/TVCG.2008.128

14. Tournier, J.D., Calamante, F., Connelly, A.: Robust determination of the fibre orientation distribution in diffusion MRI: non-negativity constrained super-resolved spherical deconvolution. Neuroimage **35**(4), 1459–1472 (2007). https://doi.org/10.1016/j.neuroimage.2007.02.016
15. Van Essen, D.C., Smith, S.M., Barch, D.M., Behrens, T.E., Yacoub, E., Ugurbil, K.: The WU-Minn human connectome project: an overview. Neuroimage **80**, 62–79 (2013). https://doi.org/10.1016/j.neuroimage.2013.05.041
16. Wakana, S., et al.: Reproducibility of quantitative tractography methods applied to cerebral white matter. Neuroimage **36**, 630–644 (2007). https://doi.org/10.1016/j.neuroimage.2007.02.049
17. Wasserthal, J., Neher, P., Maier-Hein, K.H.: Tractseg - fast and accurate white matter tract segmentation. Neuroimage **183**, 239–253 (2018). https://doi.org/10.1016/j.neuroimage.2018.07.070

Deep Multivariate Autoencoder for Capturing Complexity in Brain Structure and Behaviour Relationships

Gabriela Gómez Jiménez[✉] and Demian Wassermann[✉]

MIND, Université Paris-Saclay, Inria, CEA, Palaiseau 91120, France
`gabriela.gomez-jimenez@inria.fr`

Abstract. Diffusion MRI is a powerful tool that serves as a bridge between brain microstructure and cognition. Recent advancements in cognitive neuroscience have highlighted the persistent challenge of understanding how individual differences in brain structure influence behavior, especially in healthy people. While traditional linear models like Canonical Correlation Analysis (CCA) and Partial Least Squares (PLS) have been fundamental in this analysis, they face limitations, particularly with high-dimensional data analysis outside the training sample. To address these issues, we introduce a novel approach using deep learning- a multivariate autoencoder model-to explore the complex non-linear relationships between brain microstructure and cognitive functions. The model's architecture involves separate encoder modules for brain structure and cognitive data, with a shared decoder, facilitating the analysis of multivariate patterns across these domains. Both encoders were trained simultaneously, before the decoder, to ensure a good latent representation that captures the phenomenon. Using data from the Human Connectome Project, our study centres on the insula's role in cognitive processes. Through rigorous validation, including 5 sample analyses for out-of-sample analysis, our results demonstrate that the multivariate autoencoder model outperforms traditional methods in capturing and generalizing correlations between brain and behavior beyond the training sample. These findings underscore the potential of deep learning models to enhance our understanding of brain-behavior relationships in cognitive neuroscience, offering more accurate and comprehensive insights despite the complexities inherent in neuroimaging studies.

Keywords: diffusionMRI · Cognitive decoding · Multivariate Learning

1 Introduction

In cognitive neuroscience, a significant gap remains in understanding how interindividual differences in brain structure affect behavior [5]. Diffusion magnetic resonance imaging (dMRI) provides insights into tissue microstructure [1] enabling a more comprehensive understanding of the relationship between brain architecture and behavior. Studies like the one performed by Zimmerman et al. [3] provided evidence that grey Matter (GM) volume correlates with cognitive processes and changes across the lifespan of healthy subjects; showing that brain architecture in grey matter has a role in modulating cognition. However, linking GM microstructure to behavior is still challenging due to less established theories on GM microstructure [2].

Recent advancements in the field have shifted from one-to-one mappings between brain regions and cognition, derived from focal brain lesion studies [4], to a regional multivariate perspective [5]. Despite these advancements, challenges persist, particularly in research on healthy individuals. Most current knowledge is biased towards pathological conditions, not reflecting the complexity of brain-behavior relationships in the general population. Menon et al. [9] are among the few studies that relate GM microstructure and cognition in healthy individuals.

The Brain Structure and Behavior (BSB) community has recently emphasized two critical aspects on dMRI: techniques and methodological models [5]. From the technical perspective, the focus has primarily been on tractography and white matter, leaving grey matter and microstructure underexplored [5]. Additionally, methodologically, multivariate linear models used in BSB for dMRI data, like Canonical Correlation Analysis (CCA) and Partial Least Squares (PLS), face challenges in generalizability [5, 6]. These limitations highlight the need for more sophisticated models [5, 9, 19].

This work addresses the challenges of uncovering BSB relations in GM in two ways. By introducing an advanced neural network-based model to explore complex non-linear relationships in the multivariate

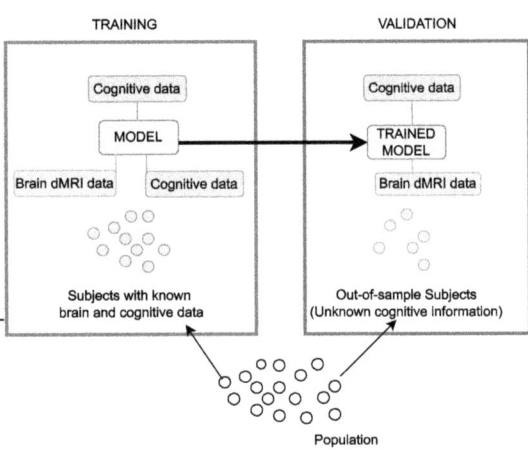

Fig. 1. Samples of populations consists of individuals with both brain data (green) and cognitive data (purple) as well as individuals with only brain data. The model is trained on the subset of subjects for whom both brain and cognitive data are available (red circles). This trained model is then applied to predict cognitive data for out-of-sample subjects, who have brain data but lack cognitive data (blue circles). The aim is to accurately predict cognitive functions from brain data, illustrating the model's capability to uncover complex BSB relationships. (Color figure online)

BSB problem, we improve generalizability. Also, we use multishell dMRI data, focusing on image attenuation from the GM of the insula, as analyzed by Menon et al. [9], to gain insights into brain microstructure. This dual approach aims to uncover deeper BSB relationships, increasing our ability to predict cognitive data from dMRI data. Figure 1 illustrates our motivation. Our goal is to develop a model capable of capturing complex BSB relationships, indicating successful identification of the necessary non-linear relations.

2 Background and Literature Review

Advancements in MRI technology have significantly influenced cognitive neuroscience, with diffusion MRI (dMRI) emerging as a powerful tool for revealing brain tissue microstructure and connectivity [1]. Traditionally focused on white matter, recent studies, such as Menon et al., show that dMRI can also provide valuable insights into grey matter (GM) microstructure and its relationship with cognitive functions [9].

2.1 Cognitive Ability Prediction in Healthy Subjects

Analyzing the diffusion signal can uncover patterns reflecting cognitive functions, advancing our understanding of brain-behavior relationships. However, the potential of dMRI to reveal cognitive processes in healthy subjects remains underexplored. Recent research, such as Porcu et al., highlights the need for further studies on the relationship between brain structure and cognition in healthy subjects, as much existing research focuses on pathological conditions [12]. While over 1,700 studies link brain structure to cognition in neurodegenerative diseases, research on healthy individuals is limited. Recent work by Kang et al. supports the consistency of GM volume studies with previous neuroimaging findings related to schizophrenia [10]. From the studies by Kang et al. and Porcu et al. [10,12], it is evident that dMRI is a powerful tool to correlate brain structure and behaviour. However, further research needs to be conducted to get more insight into the relationship between microstructure and cognition in GM. One of the relevant studies in recent years, conducted by Menon et al. [9], focused on the insula microstructure and has already provided evidence of this connection between grey matter microstructure and cognitive functions.

2.2 Multivariate Learning

While the Brain Structure and Behavior (BSB) community has several comprehensive datasets on healthy subjects, the Human Connectome Project (HCP) remains one of the most used among them. Typical neuroimaging dMRI studies face a "multivariate ill-posed" problem, where the number of subjects is significantly lower than the number of features. This affects the generalizability and stability of traditional models like Partial Least Squares (PLS) and Canonical Correlation Analysis (CCA), which tend to overfit small samples, highlighting the need for models that can handle high-dimensional data more effectively [6]. Genon et al. [5], establish PLS and CCA as the preferred approaches to unveil behavioural patterns from brain structure variables, the study conducted by Wang [7] presents the benefits of CCA as a modality fusion method between cognitive measurements and brain structure and Zhuang et al. [8] detail all the variants and ranges of CCA in this domain. Despite their limitations, linear multivariate models have been preferred in BSB research due to their ability to relate brain structure to behavioural patterns by reducing complex multivariate data into lower-dimensional representations while maintaining the phenomena' representativeness. However, these models struggle with high-dimensional dMRI data and small sample sizes, as noted by Genon et al. [5]. The inherent complexities and heteroscedasticity of real-life health data call for more advanced methodologies. Deep learning models, with their capacity to handle high-dimensional and non-linear data, offer a promising alternative. They can create an embedding space of the desired dimension, unlike CCA and PLS, whose latent space dimensions are limited by dataset features.

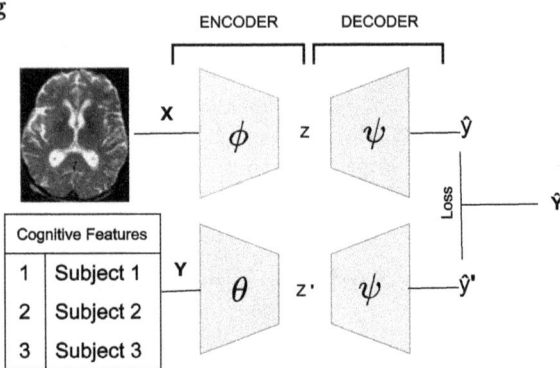

Fig. 2. The model architecture features in two encoders ϕ and θ that transform X (dMRI data) and Y (cognitive data), respectively, into lower-dimmensional embedding spaces (z and z'). These embeddings are then processed by a shared decoder ψ which reconstructs the cognitive data to produce $\hat{y}$ and $\hat{y}'$, facilitating the prediction of cognitive functions. The design allows for the exploration of complex multivariate patterns between brain structure and cognitive features.

In multivariate models, the main challenge is to effectively summarize information from multiple variables, using complementary data while filtering out redundancies [14]. Real-life data, especially health-related data, often present heteroscedasticity, meaning each variable will have different types of noise dependence, complicating the analysis. One approach to address these challenges is to align the representation of variables in a shared embedding space of reduced dimensionality, ensuring a consistent representation of similar information across different variables [13]. The advantage of deep multivariable models lies in their ability to adjust an embedding space of the desired dimension, whereas for CCA and PLS, the latent space dimension is limited by the dataset with fewer features. Given these complexities, deep learning models, with their superior capacity to handle high-dimensional and non-linear data, offer a promising alternative. In the next section, we introduce our proposed model and discuss its design and implementation in detail for BSB analysis.

3 Methodology

This study aims to develop a model that effectively predicts cognitive processes from dMRI signals while capturing the complex nature of the underlying phenomenon. We designed a multivariate Encoder-Decoder model, employing pre-processing techniques and evaluation metrics suited to the multivariate nature of the problem to ensure accuracy and generalizability. This section details the model's development and validation.

3.1 Data Preprocessing

After filtering subjects with missing information and outliers, we used the 3T dMRI and cognitive data from 779 subjects in the HCP database, focusing on the insula for its role in encoding cognitive-control functions, as noted by Menon et al. and others [9,20,21]. We selected 12 cognitive control-related processes, involving relational tasks, gambling, working memory, card and list sorting, picture sequence, flanker inhibitory control, and participant age. We standardized each cognitive feature at batch level and restored them to their original values post-prediction for consistent evaluation, given the diverse scales of the different cognitive meadurements. To define the brain structure data, and based on previous works [16–18], we hypothesize that the diffusion signal is modulated by microstructure and that microstructure has a role in modulating cognition. We performed a multishell analysis with three b-values (1000, 2000 and 3000 [s/mmÂš]). We projected the diffusion signal from the voxels in the insula onto the cortical mesh through linear interpolation to ensure consistent comparison across subjects. Using the MAP-MRI model [15], with default parameters from the dipy package, we then resampled this projected signal resulting in 9468 features per b-value, which reflects the resolution and complexity due to the numerous vertices in the cortical mesh. This resampling facilitated computation by reducing noise in the signal.

3.2 Proposed Model

Our model employs a multivariable regression Encoder-Decoder architecture with two encoder modules and a shared decoder module, as illustrated in Fig. 2.

The encoders learn joint embeddings of brain image attenuations and cognitive features first, and then, the decoder reconstructs the cognitive data from these embeddings. This sequential training approach facilitates the exploration of the complex multivariate patterns and relationships between these domains, enabling a more detailed understanding of their interactions.

Encoder Modules. The Brain Structure Encoder (ϕ) compresses brain data X into a lower-dimensional space $z(n = 64)$ using batch normalization, 5 linear layers of 6 neurons each, and LeakyReLU activation functions. Similarly, the Cognitive Data Encoder (θ) transforms cognitive data Y into a separate embedding space $z'(n = 64)$ with a comparable architecture. Both encoders, in Eq. (1), feature an input layer that accepts the respective input features and an output layer producing the final low-dimensional embeddings (z or z').

$$\phi(X) = z \qquad \theta(Y) = z'. \tag{1}$$

Shared Decoder Module. The decoder module (ψ), in Eq. (2), reconstructs cognitive data Y from embeddings z and z' ensuring a consistent mapping back to the cognitive representation. It includes the latent space input (n=64) that accepts low-dimensional embeddings (z or z'), hidden layers structured as the encoder with fully connected layers and activation functions, and an output layer (n=12) that produces reconstructed cognitive data ($\hat{y}$ and $\hat{y}'$) closely resembling the original input features.

$$\psi(z) = \hat{y} \qquad \psi(z') = \hat{y}'. \tag{2}$$

The complete Encoder-Decoder model, Fig. 2, integrates both encoders and the shared decoder into a cohesive framework, optimized using the Adam optimizer with a learning rate scheduler.

$$\begin{aligned}\psi(\phi(X) = \hat{y} \\ \psi(\theta(Y)) = \hat{y}'.\end{aligned} \tag{3}$$

3.3 Loss

The loss function measures the reconstruction error between the original and reconstructed cognitive data from the latent embeddings z and z'.

We use two different loss functions. When the encoders are training we compute $\mathcal{L}_{\text{encoders}} = \mathcal{L}(z, z')$ to ensure information from both representations is projected into the same space. Then, when the decoder is training, the total loss function $\mathcal{L}_{\text{decoder}}$ is a weighted sum of three individual loss components, all computed in the cognitive space and, therefore in the same scale:

$$\mathcal{L}_{\text{decoder}} = \alpha \mathcal{L}_{\text{embedding}} + \beta \mathcal{L}_{\text{recon\_z}} + \gamma \mathcal{L}_{\text{recon\_z'}}$$

where α, β, and γ are weight hyperparameters that can be adjusted to balance the importance of each term. They control the relative contributions of each loss component. In our analysis, all the losses had the same importance $\alpha = \beta = \gamma = 1$.

All the loss functions are the mean squared error:

$$\mathcal{L}(x,y) = \frac{1}{n}\sum_{i=1}^{n}(x-y)^2$$

where $\mathcal{L}_{\text{embedding}} = \mathcal{L}(\hat{y}, \hat{y}')$, $\mathcal{L}_{\text{recon\_z}} = \mathcal{L}(y, \hat{y})$ and $\mathcal{L}_{\text{recon\_z'}} = \mathcal{L}(y, \hat{y}')$, ensure several aspects. $\mathcal{L}_{\text{embedding}}$ establishes that the embeddings z and z' from the brain structure and cognitive data encoders are generating consistent and comparable reconstructions. The $\mathcal{L}_{\text{recon\_z}}$ focuses on reconstructing cognitive processes from brain structure embeddings z, which is our main goal. Similarly $\mathcal{L}_{\text{recon\_z'}}$ aims to reconstruct the original cognitive processes from cognitive data embeddings z', ensuring that the learned transformations from the trained encoder create an embedding space that allows accurate reconstruction of the cognitive features from the embedding space.

3.4 Validation

We implemented a cross-validation strategy with varied seed configurations, using 701 subjects for training the model and 78 subjects for validation, to evaluate the model's generalizability. We trained 5 models with different data subsets. Reconstruction accuracy was assessed using the Spearman correlation (ρ) between reconstructed and original cognitive features.

4 Results and Discussion

4.1 Cognitive Score Reconstruction

Figure 3 illustrates the cognitive reconstructions, obtained using our method presented in Sect. 3.2, derived from dMRI signal embeddings compared to original cognitive scores. This plot enables the comparison between the reconstructed and original values. Variability and heteroscedasticity in the data are reflected in deviations from the ideal diagonal line where reconstructed scores would perfectly match original scores; indicating that there is still information that the model has yet to fully capture.

The analysis across samples is presented in Table 1 where it is possible to appreciate the Spearman correlation factor between the training and evaluation steps of the model. Focusing on the Spearman correlation factor of the validation, as the interest of the paper relies on the evaluation outside the sample, it is possible to appreciate that it reached peaks around 0.2 (*Working Memory Accuracy and Reaction Time, List and Card sorting* in Fig. 3; *Working Memory Accuracy and Reaction Time* and *Processing Speed* in sample 1 Table 1; *Gambling Task* in sample 2 Table 1; *Relational Task Reaction Time* in sample 3 table 1; *List Sort* and *Age* in sample 4 Table 1); 0.29 being the highest for *List Sorting* in Fig. 3 and 0.27 for the *Working Memory Accuracy* and *Processing*

Speed in Table 1. While different subsamples learned dissimilar characteristics, complicating cross-sample comparisons, these correlations align well with significant findings in the field, such as those discussed by Menon et al. [9], where correlations around 0.2 are considered high, affirming a meaningful relationship between dMRI attenuations, and therefore brain microstructure, and behavior.

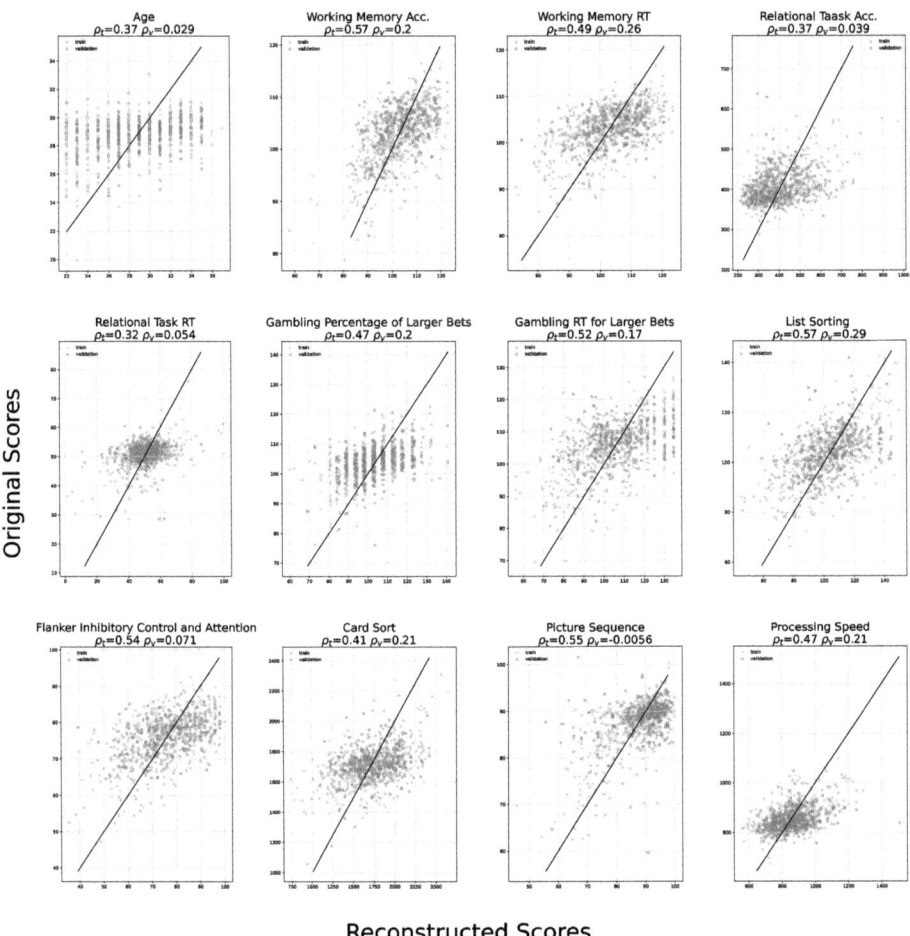

Fig. 3. Results of cognition reconstructions from dMRI attenuations represented with the respective original value. The Spearman correlation factors are shown below the name of the cognitive variable for both, training and validation processes (ρ_t and ρ_v respectively). We used 701 subjects for training and 78 for validation.

Table 1. Sample comparison of our model. Spearman's correlation factor for each cognitive feature in training and validation computed in different samples obtained from HCP with our model. This table shows the performance of our model when trained and validated on various samples.

Feature	Sample 1		Sample 2		Sample 3		Sample 4	
	Training	Validation	Training	Validation	Training	Validation	Training	Validation
Age	0.43	0.02	0.44	0.11	0.53	0.09	0.34	0.19
Working Memory Acc.	0.53	0.27	0.49	0.03	0.53	−0.03	0.47	0.11
Working Memory RT	0.62	0.22	0.43	0.03	0.63	−0.19	0.50	0.11
Relational Task Acc.	0.49	0.12	0.35	0.09	0.48	−0.01	0.44	0.15
Relational Task RT	0.40	−0.06	0.31	0.16	0.51	−0.24	0.39	0.02
Gambling Task	0.49	0.19	0.40	0.24	0.39	−0.12	0.47	0.13
Gambling Task RT	0.57	0.17	0.34	0.02	0.43	0.07	0.51	0.03
List Sorting	0.55	0.00	0.41	0.04	0.47	0.11	0.45	−0.20
Flanker	0.60	0.06	0.34	0.07	0.55	0.03	0.50	0.07
Card Sorting	0.51	0.01	0.57	0.00	0.53	0.08	0.40	0.07
Picture Sequence	0.57	0.12	0.50	0.08	0.47	0.07	0.50	0.02
Processing Speed	0.57	0.27	0.55	0.11	0.52	0.07	0.44	0.01

4.2 Model Comparison

To contextualize our results, we compared our model's performance with traditional methods: CCA and PLS. These metrics were computed similarly to our multimodal learning model: normalizing training data and reconstructing predictions using mean and standard deviation derived from the train set. We performed a $k = 5$ k-fold cross-validation as a baseline for both CCA and PLS to ensure robust model evaluation. These results are illustrated in Table 2 alongside our model's best results obtained.

Table 2 presents Spearman correlation values for both training and validation phases and our analysis focuses on two key aspects.

First, in the validation outcomes, as this assessment underscores the ability to capture the correlation between $\hat{y}$ and y across various cognitive variables not seen during training. Our model outperforms CCA and PLS in capturing these correlations, showing a better performance in predicting multivariate cognitive processes. Second, the training and validation gap, as it provides insights into the model's performance and generalization. While our model shows better results compared to CCA and PLS, the larger gap between training and validation suggests potential overfitting. However, in the context of complex phenomena such as BSB, our primary focus is on achieving good out-of-sample performance rather than minimizing the gap between training and validation set performance. If the validation set represents the data the model will encounter in practice, good validation performance suggests that overfitting is not an issue, regardless of the observed gap. Notably, in this type of complex problem, reconstructions with a correlation of 0.2 are considered good [9].

Table 2. Model comparison. Spearman's correlation factor for each cognitive feature in training and validation using CCA, PLS, and our model, based in multimodal learning.

Feature	CCA		PLS		Ours	
	Training	Validation	Training	Validation	Training	Validation
Age	**1.00**	**0.17**	0.22	0.13	0.37	0.03
Working Memory Acc.	0.45	0.14	0.47	0.16	**0.57**	**0.20**
Working Memory RT	0.24	0.14	0.32	0.06	**0.49**	**0.26**
Relational Task Acc.	0.34	**0.16**	0.45	0.01	**0.37**	0.039
Relational Task RT	0.25	**0.15**	0.15	0.11	**0.32**	0.054
Gambling Task	0.30	0.15	0.09	0.09	**0.47**	**0.20**
Gambling Task RT	0.11	0.03	0.25	0.00	**0.52**	**0.17**
List Sorting	0.17	−0.05	0.24	0.16	**0.57**	**0.29**
Flanker	0.03	0.06	0.35	0.07	**0.54**	**0.07**
Card Sorting	0.37	0.12	**0.47**	0.15	0.41	**0.21**
Picture Sequence	0.19	0.04	0.16	**0.19**	**0.55**	−0.01
Processing Speed	0.20	0.04	0.32	0.06	**0.47**	**0.21**

Nevertheless, we can appreciate that our model faces challenges in generalizing across different samples, as some of the reconstruction scores of the cognitive features vary depending on the samples as it is possible to appreciate in Table 1, in features like *Working Memory Accuracy* and *Relational Task Reaction Time*. This underscores the need for further research and refinement of the multimodal architecture. Addressing these challenges is crucial, especially given the noise and variability when integrating various data modalities in such complex problems.

5 Conclusions

In this study, we addressed the challenge of understanding the relationship between diffusion MRI attenuations in grey matter, an intermediary for brain structure [15–17], and cognitive functions by employing a novel neural network-based approach. Traditional linear models such as CCA and PLS have shown limitations, particularly in terms of their generalizability and stability when applied to high-dimensional data from subjects outside of the sample.

Our multimodal learning model, which employs distinct encoder modules for brain structure and cognitive data and a shared decoder module, significantly improved the ability to capture correlations between these domains, as shown in Table 2. The validation Spearman coefficient indicates a better performance compared to state-of-the-art methods. Although a gap between training and validation suggests potential overfitting, this observation should be interpreted with caution due to the complexity of the phenomenon. As shown in Table 2, for the *List Sorting* cognitive feature, our model exhibits a training-validation gap of 0.28, compared to a lower gap of 0.08 in the PLS model. However, despite the larger gap, our model achieves a superior reconstruction score in out-of-sample data. This gap along with the varied reconstruction of cognitive variables shown in Table 1, implies that cognitive variables might not be fully independent, as noted by Menon [9]. Therefore, further exploration in a latent space may be necessary.

It is important to emphasize that this study is not focused on interpretable machine learning for the BSB phenomenon. We consider that the diffusion signal, modulated by microstructure, reflects a correlation between cognition and microstructure through its relationship with the signal. While further work is needed to ensure robustness and interpretability, our model offers a more flexible starting point compared to classical approaches.

These findings suggest that complex, non-linear models can provide deeper insights into brain-behavior relationships, offering a more accurate and comprehensive understanding of cognitive neuroscience. This is particularly relevant given the challenges traditional models face in handling high-dimensional neuroimaging data and its inherent variability. Our study contributes to the growing body of evidence supporting the promise of deep models in advancing cognitive neuroscience, especially when dealing with diffusion data in grey matter to uncover complex relationships with cognitive features in healthy subjects. Future research should not only explore integrating additional data dimensions, such as structural representations, but also consider the use of latent space models. Investigating latent space could reveal underlying patterns and relationships that are not captured by conventional approaches, potentially increasing our understanding of the cognitive processes and their neural base. This approach may further improve model performance and provide more insights into the brain-behavior intersection.

Acknowledgements. We thank the Human Connectome Project [22] for providing the data used in this study. The data were anonymized according to the stringent protocols established by the HCP.

We also wish to acknowledge the support of the ANR Project MicBrainPres and the Jean Zay cluster for their essential contributions to this research. In particular, we appreciate the access granted to the HPC resources of IDRIS under the allocation 2024-AD01101484 made by GENCI, which was instrumental in carrying out our computations.

References

1. Le Bihan, D., Iima, M.: Diffusion magnetic resonance imaging: what water tells us about biological tissues. PLoS Biol. **13**(7), e1002203 (2015)
2. Novikov, D.S., Kiselev, V.G., Jespersen, S.N.: On modeling. Magn. Reson. Med. **79**(6), 3172–3193 (2018)
3. Zimmerman, M.E., et al.: The relationship between frontal grey matter volume and cognition varies across the healthy adult lifespan. Am. J. Geriatric Psychiatry **14**(10), 823–833 (2006)

4. Genon, S., Reid, A., Langner, R., Amunts, K., Eickhoff, S.B.: How to characterize the function of a brain region. Trends Cogn. Sci. **22**(4), 350–364 (2018)
5. Genon, S., Eickhoff, S.B., Kharabian, S.: Linking interindividual variability in brain structure to behaviour. Nat. Rev. Neurosci. **23**(5), 307–318 (2022)
6. Helmer, M., et al.: On the stability of canonical correlation analysis and partial least squares with application to brain-behavior associations. Commun. Biol. **7**(1), 217 (2024)
7. Wang, H.T., Smallwood, J., Mourao-Miranda, J., Xia, C.H., Satterthwaite, T.D., Bassett, D.S., Bzdok, D.: Finding the needle in a high-dimensional haystack: canonical correlation analysis for neuroscientists. Neuroimage **216**, 116745 (2020)
8. Zhuang, X., Yang, Z., Cordes, D.: A technical review of canonical correlation analysis for neuroscience applications. Hum. Brain Mapp. **41**(13), 3807–3833 (2020)
9. Menon, V., et al.: Microstructural organization of human insula is linked to its macrofunctional circuitry and predicts cognitive control. eLife **9**, e53470 (2020)
10. Kang, N., Chung, S., Lee, S.H., Bang, M.: Cerebro-cerebellar grey matter abnormalities associated with cognitive impairment in patients with recent-onset and chronic schizophrenia. Schizophrenia **10**(1), 11 (2024)
11. Zhu, A. H., et al.: Lifespan reference curves for harmonizing multi-site regional brain white matter metrics from diffusion MRI. In: bioRxiv 2024-02 (2024)
12. Porcu, M., Cocco, L., Marrosu, F., Cau, R., Suri, J.S., Qi, Y., Saba, L.: Impact of corpus callosum integrity on functional interhemispheric connectivity and cognition in healthy subjects. Brain Imaging Behav. **18**(1), 141–158 (2024). https://doi.org/10.1007/s11682-022-00472-9
13. Andrew, G., Arora, R., Bilmes, J., Livescu, K.: Deep canonical correlation analysis. In: Dasgupta, S., McAllester, D. (eds.) International Conference on Machine Learning, pp. 1247–1255. PMLR (2013)
14. Friston, K.J.: Eigenimages and multivariate analyses. In: Frackowiak, R.S.J., Friston, K.J., Frith, C.D., Dolan, R.J., Mazziotta, J.C. (eds.) Human Brain Function (1997)
15. Özarslan, E., Koay, C.G., Shepherd, T.M., Komlosh, M.E., İrfanoğlu, M.O., Pierpaoli, C., Basser, P.J.: Mean apparent propagator (MAP) MRI: a novel diffusion imaging method for mapping tissue microstructure. Neuroimage **78**, 16–32 (2013)
16. Afzali, M., et al.: The sensitivity of diffusion MRI to microstructural properties and experimental factors. J. Neurosci. Methods **347**, 108951 (2021)
17. Avram, A.V., et al.: Clinical feasibility of using mean apparent propagator (MAP) MRI to characterize brain tissue microstructure. Neuroimage **127**, 422–434 (2016)
18. Fick, R.H., Wassermann, D., Caruyer, E., Deriche, R.: MAPL: tissue microstructure estimation using Laplacian-regularized MAP-MRI and its application to HCP data. Neuroimage **134**, 365–385 (2016)
19. Mihalik, A., et al.: Canonical correlation analysis and partial least squares for identifying brain–behavior associations: a tutorial and a comparative study. Biol. Psychiat. Cogn. Neurosci. Neuroimaging **7**(11), 1055–1067 (2022)
20. Ham, T., Leff, A., de Boissezon, X., Joffe, A., Sharp, D.J.: Cognitive control and the salience network: an investigation of error processing and effective connectivity. J. Neurosci. **33**(16), 7091–7098 (2013)
21. Molnar-Szakacs, I., Uddin, L.Q.: Anterior insula as a gatekeeper of executive control. Neurosci. Biobehav. Rev. **139**, 104736 (2022)
22. Van Essen, D.C., et al.: The Human Connectome Project: a data acquisition perspective. Neuroimage **62**(4), 2222–2231 (2012)

Heritability and Genetic Correlations Along the Corticospinal Tract

Iyad Ba Gari[1](✉), Ravi R. Bhatt[1], Fang-Chang Yeh[2], and Neda Jahanshad[1]

[1] Laboratory of Brain eScience, Mark and Mary Stevens Neuroimaging and Informatics Institute, Keck School of Medicine, University of Southern California, Marina del Rey, CA, USA
{bagari,njahansh}@usc.edu
[2] Department of Neurological Surgery, University of Pittsburgh, Pittsburgh, USA

Abstract. Diffusion-weighted magnetic resonance imaging (dMRI) allows for the mapping and analysis of white matter (WM) microstructure and structural connectivity in the brain. We describe and extend the Medial Tractography Analysis (MeTA) method to capture the regional along-tract variation of WM microstructure by incorporating hyperplanes to capture curvature related to cortical connections. We performed heritability and genetic correlation analysis of the fractional anisotropy (FA) of the left and right corticospinal tract (CST) along its length using data from 20,734 participants in the UK Biobank. We found that segment-specific heritability (h^2_{SNP}) within the CST varied specific segments and showed higher heritability than the average FA across the full CST. Genetic and phenotypic correlations along CST segments of FA reveal moderate to high correlations and suggest shared genetic influences on FA across segments with many strong lateralized correlations. Using hierarchical clustering, we identified two primary genetically correlated clusters along the CST. These findings indicate homogeneous genetic factors within each cluster and regional variability in genetic influences across clusters. Our source code is available at https://github.com/USC-LoBeS/MeTA.

Keywords: Diffusion MRI · Heritability · Genetic Correlation · Corticospinal Tract · Along Bundle Analysis

1 Introduction

Diffusion-weighted magnetic resonance imaging (dMRI) allows us to map and analyze the microstructural characteristics of white matter (WM) and structural connectivity in the brain. D-MRI can be used to generate whole-brain tractograms, which can be segmented into WM bundles or pathways [28]. The microstructural profile along these pathways can be estimated via the diffusion tensor imaging (DTI) among other model using metrics such as fractional anisotropy (FA). WM bundles provide virtual insights into the normal and pathological variations within the WM, enabling studies on brain development, and neurological and psychiatric disorders across the lifespan [15,18].

Individual differences in white matter microstructure among the general population are significantly influenced by genetic variation. This knowledge provides a crucial link between our genes, brain function and connectivity, offering valuable insights into the development of neuropsychiatric disorders [7]. Zhao et al. showed that FA in the global average across 21 tracts and several white matter bundles had higher SNP heritability compared to DTI metrics such as AD, RD, MD, and MO. Population-based studies have demonstrated that the average FA of the entire corticospinal tract (CST) in both hemispheres exhibits low to moderate heritability compared to other cerebral white matter bundles. Using twin and pedigree data, the heritability estimate for the average bilateral skeletonized CST FA has been reported to be approximately 66% [13]. A recent study using genome-wide information from the UK Biobank dataset ($N = 34{,}024$) showed a heritability of average FA in the CST was approximately 30% using genome-wide complex trait analysis tool (GCTA-GREML) [26,34], substantially lower than other WM regions of interest. Elliott et al. used data from the UK Biobank ($N = 7{,}532$) to identify genetic variants associated with the CST FA, and estimate its heritability, using two methods: tract-based spatial statistics (TBSS) and probabilistic tracking with crossing fibres (ProbtrackX). The estimated heritability for CST using TBSS was 22–23%, and 33–35% using ProbtrackX [4,8,23], highlighting how tractography based methods capture more of the genetic variance. However, more sensitive tractography based approaches may explain greater variability and capture nuanced local genetic influences along the tract, instead of disregarding regional variations by averaging FA along the entire tract.

Different WM bundle segmentation tools and methods have been shown to affect the measurements of DTI metrics, as detailed in Schilling et al. [20]. Several approaches perform population-based analyses of microstructure along WM bundles. For example, Bundle Analytics (BUAN) is a tractometry tool that analyzes all possible streamlines within a bundle to incorporate anatomical variations [6]. However, the streamline-based BUAN may require substantial computational resources for imaging genetics, as the computation time increases with the number of streamlines and the dataset's sample size. As imaging genetic studies require large sample sizes, faster and scalable methods are needed. Medial Tractography Analysis (MeTA) has been developed to analyze the WM bundle along its length of the volume, rather than on streamlines. MeTA extracts and parcellates core volumes of tractography defined WM bundles along their length directly in voxel space, preserving bundle shape and capturing regional variations of the WM tracts [2,3].

In this work, we describe and extend the MeTA approach to segment the CST bundle using hyperplanes, which takes into account the fanning and splitting of the CST architecture, rather than using a straight plane. We used MeTA to study the genetic architecture along the CST and we investigated the heritability, phenotypic and genetic correlations of FA along 15 segments of the CST in the

UK Biobank dataset [5], providing rationale for how the number of segments was selected. We use hierarchical clustering methods to identify segments along the CST that have relatively homogeneous genetic architecture.

2 Methods

2.1 MeTA: Medial Surface Generation

For a 3D object shape X, a Continuous Medial Representation (CM-Rep) m is defined as a parameterized continuous medial surface model with a radius scalar field R [32,33]. The shape X is reconstructed using a maximum inscribed ball (MIB) with radius R at each point on m. The shape X can be parameterized as $X^{\pm}(u)$, where X^+ and X^- are on opposite sides of the parameterized medial surface $m(u)$ with u being a surface parameter of m.

$$X^{\pm}(u) = m(u) + R(u)U^{\pm}(u) \tag{1}$$

where, $U^{\pm}$ are the unit outward normal vectors on either side of the medial surface defined as:

$$U^{\pm} = -\nabla_m R \pm \sqrt{1 - \|\nabla_m R\|^2}\, N_m \tag{2}$$

Here, N_m is the unit normal vector to the medial surface $m(u)$ and $\nabla_m R$ is the gradient of the radius field with respect to the medial surface. The CST example is shown in Fig. 1.

2.2 MeTA: Medial Core Extraction

We calculated the normals at the vertex points on the medial surface (m) and found the corresponding points on the boundary mesh $X^{\pm}$ by traveling along the normal vector in both directions using ray tracing methods [24] which involve casting rays from the vertex points on the medial surface (m) along the normal vectors to determine where they intersect with the boundary mesh ($X^{\pm}$).

$$P^{\pm}(u) = m(u) + \lambda N(u) \tag{3}$$

where $P^{\pm}(u)$ are the points on the boundary mesh reached by traveling along the normals of $m(u)$, the medial surface. $N(u)$ is the normal vector at $m(u)$ and λ is the percentage distance traveled along the normal vector from $m(u)$ to each the boundary mesh ($X^{\pm}$). We computed the distance (thickness/depth) from each vertex point on the medial surface $m(u)$ to its closest vertex point on the outer boundary mesh $X^{\pm}$ that lies along the normal axis in both directions using the k-dimensional tree (KD-Tree) method which speeds up the computation of distances [21].

$$D_{\min} = \min(\|P^+(u) - m(u)\|, \|P^-(u) - m(u)\|) \tag{4}$$

where $D_{\min}$ is the shortest distance, and $P^+(u)$ and $P^-(u)$ are the points on the boundary mesh in the positive and negative directions of the normal vector $N(u)$, respectively.

To obtain the central 25% 'core' of the 3D volumetric shape, we computed the 37.5% and 62.5% surfaces by setting $\lambda = 12.5\%$ (12.5% around the medial 50% surface) and then voxelized these surfaces to extract the sub-volume (MeTA 25%) from the original boundary X as shown in Fig. 1 [3].

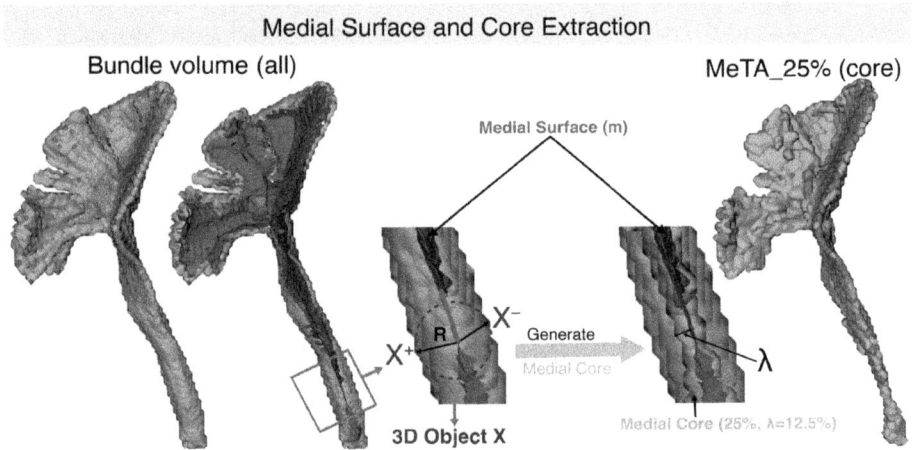

Fig. 1. Medial surface and core extraction workflow. The generated corticospinal tract (CST) bundle from DSI-Studio converted to the volume of interest. The *green color* illustrates the constructed boundary object, while *magenta* represents the medial surface of the object. The *orange color* shows the medial core of the CST (MeTA_25% where $\lambda = 12.5\%$); this percentage indicates the distance traveled along the normal vector from the medial surface (*magenta*) to each boundary mesh (X) in *green*. (Color figure online)

2.3 MeTA: Bundle Core Parcellation

We extended MeTA by developing a hyperplane parcellation along the bundle core (MeTA_25% 'core'). This is done by generating a centroid along the length of the template bundle and dividing it into P points for reference in cross-subject comparisons to establish correspondence between subjects [2,29]. For each subject's bundle, a centroid was generated with 500 points using QuickBundles [10]. Dynamic time warping (DTW) was then applied to align corresponding points between the subject and template bundles [19]. DTW optimizes the correspondence between two sets of points by minimizing the Euclidean distance between the two sequences under all possible sets of alignments. Consequently, each point in the template's centroid is matched with multiple points in the subject's centroid, from which we select the midpoint and return the corresponding P points

(15 points as used for bundle segmentation). We generated centroids from the subject's bundle using QuickBundles clustering and then established corresponding points within P points of the subject's centroid and these generated centroids for better bundle shape representation. We combined all corresponding points to provide a more robust representation of the WM bundle and ensure consistency across different subjects in parcellating the MeTA_25% 'core'.

Using these corresponding points (P), we divided the full 3D binary mask of the bundle ('all') for each subject into multiple segments as illustrated in Fig. 2. Each voxel of the bundle was assigned to a segment based on its relative position to the DTW points P, determined using dot product ($\cdot$) calculations between vectors of consecutive corresponding points and vectors of the current corresponding point with all other points. The details of bundle core parcellation along its length into N segments, where P is the number of centroid points, are as follows:

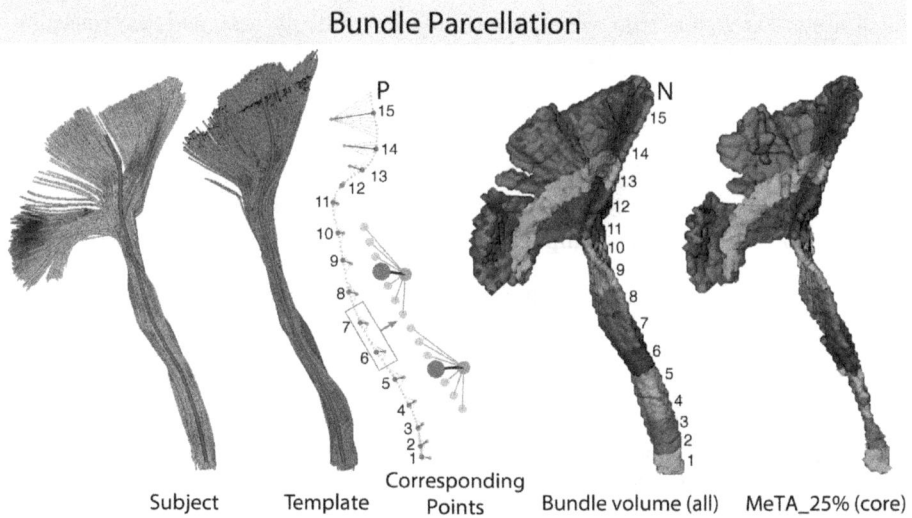

Fig. 2. Bundle core parcellation workflow. The generated centroid, shown in *red*, along the length of the DSI-Studio template in *blue* was divided into $P = 15$ points as a reference in cross-subject comparison. A centroid in *red* was generated with 500 points for each subject (*orange*) using QuickBundles [10]. Dynamic time warping (DTW) was used to get corresponding points by aligning these two centroids. The CST core volumes were parcellated into N = 15 segments using corresponding points. (Color figure online)

- First, we calculated the normal vector ($n_1 = P_2 - P_1$) for the first segment ($i = 1$) and iterated for each voxel with coordinates V in the bundle binary mask:

$$if\ (V - P_1) \cdot n_1 \geq 0,\ then\ S(V) = 1 \tag{5}$$

- For each intermediate segment ($1 < i < N-1$), we computed the normal vectors for current ($n_i = P_{i+1} - P_i$) and subsequent planes ($n_{i+1} = P_{i+2} - P_{i+1}$) and iterated for each voxel V in the bundle's binary mask:

$$if\ (V - P_i) \cdot n_i \geq 0\ and\ (V - P_{i+1}) \cdot (-n_{i+1}) \geq 0,\ then\ S(V) = i+1 \quad (6)$$

- For the second to last segment ($i = N-1$), we calculated the normal vector ($n_{N-1} = P_N - P_{N-1}$) for the second to last segment and iterated for each voxel V in the bundle binary mask:

$$if\ (V - P_{N-1}) \cdot n_{N-1} \geq 0,\ then\ S(V) = N - 1 \quad (7)$$

- Finally, we calculated the normal vector ($n_N = P_N - P_{N-1}$) for the last segment ($i = N$) and iterated for each voxel V in the bundle binary mask:

$$if\ (V - P_N) \cdot n_N \geq 0,\ then\ S(V) = N \quad (8)$$

For ambiguous voxels that were unassigned or multiply assigned, we reassigned them to the segment with the closest DTW point according to the minimum Euclidean distance.

$$S(V) = \arg\min_i \|V - P_i\| \quad (9)$$

where, V is the voxel position and P_i are DTW points. We selected the total number of segments (N) based on the following equation to provide a standard reference for cross-subject comparison:

$$N = \frac{\text{Average bundle length (mm)}}{\text{voxel size (mm)} \times \text{required thickness}} \quad (10)$$

We chose a required thickness of four voxels to maintain consistency across different WM bundles and to ensure that segments are sufficiently large and not fragmented, as WM bundle length varies across subjects. P was also set to the same number of N. Code for MeTA is publicly available at https://github.com/USC-LoBeS/MeTA.

2.4 Dataset

Diffusion MRI data from 20,734 participants from the UK Biobank (UKB) were used in our analysis [5]. These participants were of European genetic ancestry (according to data field 22006 of the UKB) and aged between 45–82 years (female = 53%). Processed dMRI scans and FA were downloaded from data field 20250 of the UKB. These processed dMRI scans had been corrected for eddy current and motion artifacts as previously described [1]. We reconstructed the corticospinal tract (CST) using DSI studio software (http://dsi-studio.labsolver.org), as follows: dMRI images were reconstructed using generalized q-sampling imaging with a diffusion sampling length ratio of 1.25 [31]; restricted diffusion was quantified using restricted diffusion imaging; automatic fiber tracking was used to reconstruct the CST bundle using deterministic fiber tracking with augmented tracking strategies, such as bundle length constraints determined from the DSI Studio atlas and pruning iterations set to zero, to improve reproducibility [28,30]. A density map of CST bundle was then created from the streamlines and then converted into a binary CST mask with a threshold equal to zero. We applied the medial tractography analysis (MeTA) approach to segment the binary mask of the CST along its length into 15 segments across each hemisphere based on Eq. 10, where the average CST length was 121.5 mm and the voxel size was 2 mm. Then, we computed the mean of FA across all voxels within the parcel.

In our analysis, we defined 33 features, including the core of the left and right CST bundle, their 15 bilateral segments along the core length, and whole-brain white matter, which was extracted using FreeSurfer 7.1 [9]. We estimated the phenotypic correlations by converting t-values from multiple linear regressions to partial correlations, adjusting for covariates including age, sex, the interaction between age and sex, total intracranial volume, scanner site, and the first 10 genotypic principal components. We estimated SNP-based heritability h^2_{SNP} for the 33 traits using the GREML approach implemented in the Genome-wide Complex Trait Analysis (GCTA) tool [26,27], adjusting for the same covariates as in the phenotypic correlations. The SNP-based genetic correlations (r_g) between 33 traits were estimated using bivariate GREML analysis implemented in GCTA [16,27], adjusting for the same covariates as in the phenotypic correlations. In the genetic correlation analysis, we tested the hypotheses of significance by fixing the genetic correlation at zero and one using a log-likelihood test, to test for a significant genetic correlation different from zero ($r_g \neq 0$), and differences in genetic composition ($r_g \neq 1$) respectively. The significance of the heritability estimates was determined using the false discovery rate (FDR) procedure, with a threshold of q = 0.05, to account for multiple comparisons across 33 regions. Genotypic correlations were corrected for multiple comparisons across 528 tests, and phenotypic correlations were separately corrected for 528 tests, using the FDR procedure with a threshold of q = 0.05. We used phenotype and genetic correlations as input for hierarchical clustering. The hierarchical clustering used Euclidean distance to compute the initial pairwise distances between individual data points of phenotype and genotype correlation and used the Ward variance minimization algorithm to compute the distances between clusters [17]. The clustering results were visualized using a dendrogram as shown in Fig. 5.

3 Results

We estimated the heritability of FA in the white matter of the brain, the core of the CST bundle, and 15 segments along its length for each of the left and right hemispheres, and we observed significant genetic influences across these regions. The overall WM had significant SNP heritability (h^2_{SNP} = 54.03%, SE= 0.03). In the core of the CST, the heritability was estimated at 34% for the left side and 32% for the right side. Using MeTA, we examined the heritability across 15 segments of the CST core. Segment 10 displayed the highest heritability (h^2_{SNP} = 45% on the right and 44% on the left brain hemispheres), followed by segment 11, with $h^2_{SNP} = 43\%$ for both brain hemispheres as shown in Fig. 3A and 4.

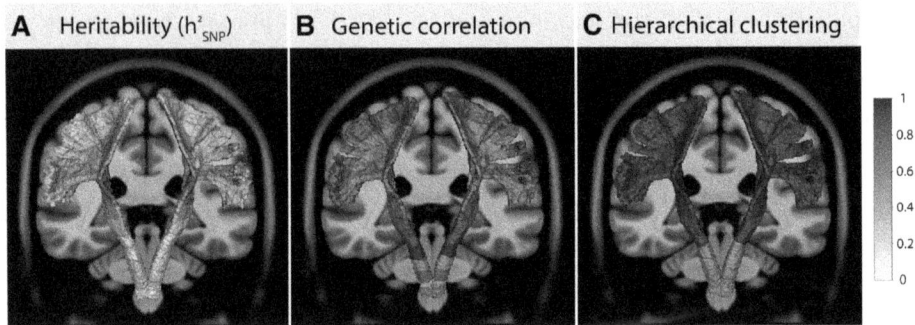

Fig. 3. A. Heritability h^2_{SNP} along 15 bilateral segments of the corticospinal tract (CST) estimated using the GREML method [26] implemented in GCTA [27] and adjusted for age, sex, age*sex interaction, total intracranial volume, scanner site, and the first 10 genotype principal components. **B.** The genetic correlation along 15 bilateral segments of the CST was estimated using Bivariate GREML analysis [16] implemented in GCTA [27] and adjusted for the same covariates as heritability. Uniquely colored segments indicate genetic correlation where the log-likelihood test is significant with respect to zero and not significant with respect to one, suggesting significant heritability and genetic correlations that are not significantly different than 1 (uniquely colored segments are not statistically different from each other). The gray color indicates no significant genetic correlations between left and right segments. **C.** Hierarchical clustering of the genetic correlations along 15 bilateral segments of the CST two resulted in major clusters using the Ward method. The first cluster includes segments one through five, while the remaining segments are grouped into the second cluster.

The analysis of phenotypic correlations across the CST segments revealed distinct patterns of association between segments in the right and left hemispheres. We found a moderate positive partial phenotypic correlation of FA between these 15 segments. We identified a highly positive genetic correlation along the CST across these 15 segments, as illustrated in Fig. 4. We also found a negative correlation between the beginning and ending segments within the CST and across hemispheres. For example, the first (most inferior) segment of the left CST negatively correlated with the 11th segment in the right CST, and the second segment of the left CST negatively correlated with the 10th segment in the left CST, as shown in Fig. 4. We identified several traits that overlap and have indistinguishable genetic factors as shown in Fig. 3B. These were largely restricted to the same segments across hemispheres.

In the hierarchical clustering analysis, phenotypic and genotypic correlations of FA along 15 segments of the CST core were grouped into two primary clusters for each correlation, as shown in Fig. 5A and 5B. For phenotypic correlations,

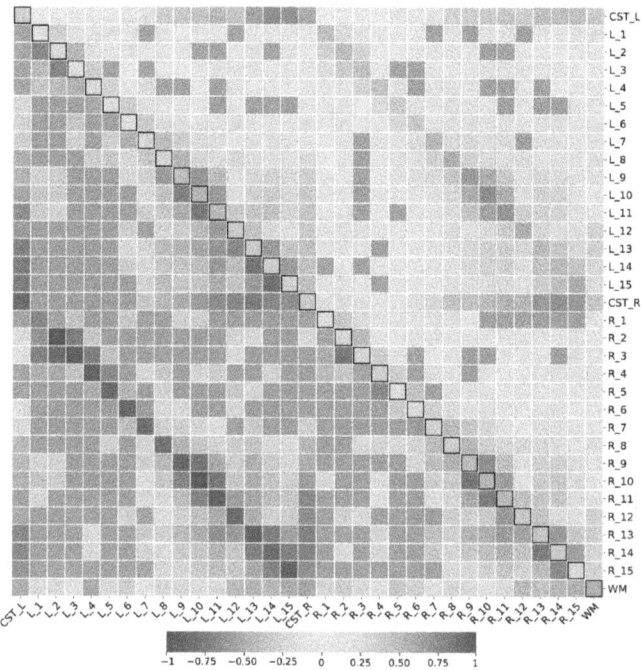

Fig. 4. A Correlation matrix of 15 segments of the corticospinal tract (CST) and white matter (WM) of the brain. The upper triangle represents phenotype correlations, the diagonal shows heritability, and the lower triangle indicates genetic correlations. Gray color indicates no significant association, while crossed (X) markers represent genetic correlations where the log-likelihood test is not significant with respect to one. (Color figure online)

cluster 1 includes segments 1 through 7, while cluster 2 comprises segments 8 through 15 from both the left and right brain hemispheres. Similarly, cluster 1 of the genotypic correlation includes segments 1 to 5, while cluster 2 includes segments 6 through 15 as shown in Fig. 3C.

4 Discussion and Conclusions

In this study, we proposed a comprehensive analysis of the heritability and genetic correlation of FA in the WM of the brain, with a particular focus on the length of the CST across 15 segments. The significant SNP heritability observed across these regions underscores the genetic influence on white matter microstructure, which has important implications for understanding neurological and psychiatric conditions [34]. We quantified significant genetic influences throughout the brain white matter, with a SNP heritability estimate of 54.03% for overall white matter, indicating a strong genetic basis for microstructural differences in the WM region. The CST core showed moderate heritability, with

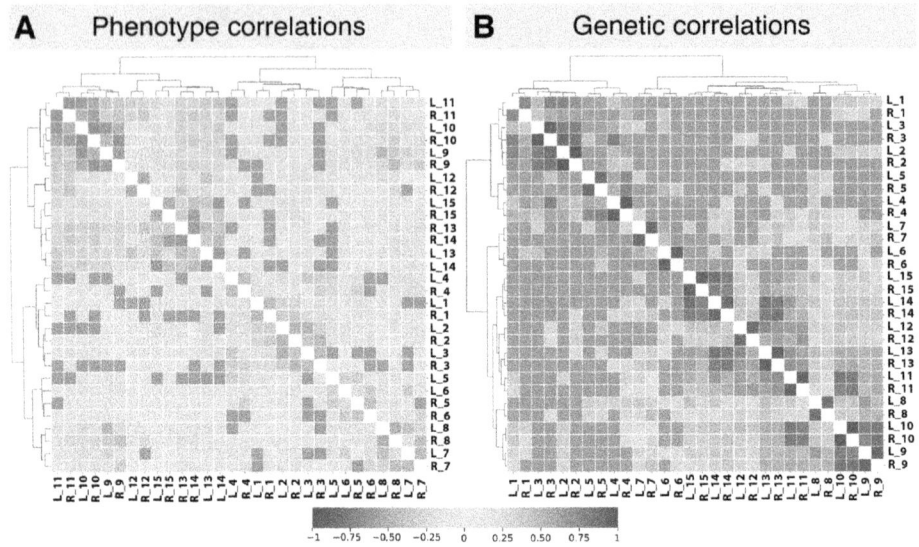

Fig. 5. A. The dendrogram was generated from the hierarchical clustering based on phenotype correlations among clusters and across 15 segments of the corticospinal tract (CST). A heatmap indicates significant associations of the phenotype correlations that pass the FDR threshold ($P_{FDR} < 0.05$), while the gray color indicates no significant association. **B.** The dendrogram was derived from the hierarchical clustering based on genetic correlations among clusters and across 15 segments of the CST. A heatmap indicates significant associations of the genetic correlations within and between clusters that pass the FDR threshold ($P_{FDR} < 0.05$). The gray color indicates no significant association and crossed (X) markers represent genetic correlations where the log-likelihood test is not significant with respect to one.

estimates of 34% for the left side and 32% for the right. This finding aligns with previous studies that have highlighted the strong genetic component in the CST structure and integrity [14,25,34]. To the best of our knowledge, our study is the first study analyzing heritability along the length of CST core across multiple segments in a large population. Our segment-specific analysis of the CST revealed the variability in heritability estimates along its length, with highly symmetric genetic influences. Segment 10 displayed the highest heritability (45% on the right and 44% on the left brain hemispheres), followed by segment 11 with heritability values of 43% for both brain hemispheres. Crossing fibers occur in these regions, where several bundles including the corpus callosum, superior longitudinal fasciculus, corona radiata, and cingulum bundle interact with the CST. [11]. The high heritability may indicate a genetic contribution to the crossing driving trait variability. These findings suggest that the genetic contribution to the microstructure of the CST is not uniform along its entire length, and are congruent with high heritability of commissural fibers which cross through this segment shown in the whole-brain WM connectome [22]. The clustering above and below the pons due to distinct genetic influences is likely highly influenced by the convergence of fibers of the corticobulbar tract - providing input to nuclei and cranial nerves innervating the face, head and neck - and CST in the brainstem [12]. The analysis of phenotypic and genotypic correlations of the CST core across 15 segments showed moderate to high correlation between these segments, indicating shared genetic factors that influence the FA of the CST. Several pairs of traits with strong genetic overlap were identified, as represented in Fig. 4 with a cross symbol. We performed hierarchical clustering analysis for phenotypic and genotypic correlations of FA along the CST core segments and identified two primary clusters for each correlation type. This clustering suggests that there are relatively homogeneous genetic factors between segments within each cluster, and that the genetic influences on the CST are largely symmetric.

The MeTA approach allows us to capture the regional heritability and genetic correlation within the core of CST. Future work will involve analyzing more white matter bundles and performing genome-wide association studies (GWAS) along the bundles' length.

Acknowledgments. This research was conducted using the UK Biobank Resource under Application Number "11559", and was supported in part by NIH grants R01MH134004, RF1NS136995, P41EB015922, S10OD032285 and NSF GRFP 2020290241.

References

1. Alfaro-Almagro, F., et al.: Image processing and quality control for the first 10,000 brain imaging datasets from UK biobank. Neuroimage **166**, 400–424 (2018)
2. Ba Gari, I., et al.: Along-tract parameterization of white matter microstructure using medial tractography analysis (MeTA). In: The 19th International Symposium on Medical Information Processing and Analysis (2023)
3. Ba Gari, I., et al.: Medial tractography analysis (MeTA) for white matter population analyses across datasets. In: 2023 11th International IEEE/EMBS Conference on Neural Engineering (NER), pp. 1–5, April 2023
4. Behrens, T.E.J., et al.: Probabilistic diffusion tractography with multiple fibre orientations: What can we gain? Neuroimage **34**(1), 144–155 (2007)
5. Bycroft, C., et al.: The UK biobank resource with deep phenotyping and genomic data. Nature **562**(7726), 203–209 (2018)
6. Chandio, B.Q., et al.: Bundle analytics, a computational framework for investigating the shapes and profiles of brain pathways across populations. Sci. Rep. **10**(1), 17149 (2020)
7. Chen, N., et al.: Meta-analyses of RELN variants in neuropsychiatric disorders. Behav. Brain Res. **332**, 110–119 (2017)
8. Elliott, L.T., et al.: Genome-wide association studies of brain imaging phenotypes in UK biobank. Nature **562**(7726), 210–216 (2018)
9. Fischl, B.: FreeSurfer. Neuroimage **62**(2), 774–781 (2012)
10. Garyfallidis, E., et al.: QuickBundles, a method for tractography simplification. Front. Neurosci. **6**, 175 (2012)

11. Glenn, G.R., et al.: Mapping the orientation of white matter fiber bundles: a comparative study of diffusion tensor imaging, diffusional kurtosis imaging, and diffusion spectrum imaging. AJNR Am. J. Neuroradiol. **37**(7), 1216–1222 (2016)
12. He, J., et al.: Reconstructing the somatotopic organization of the corticospinal tract remains a challenge for modern tractography methods. Hum, Brain Mapp (2023)
13. Kochunov, P., et al.: Heritability of fractional anisotropy in human white matter: a comparison of human connectome project and ENIGMA-DTI data. Neuroimage **111**, 300–311 (2015)
14. Kochunov, P., et al.: Multi-site study of additive genetic effects on fractional anisotropy of cerebral white matter: Comparing meta and megaanalytical approaches for data pooling. Neuroimage **95**, 136–150 (2014)
15. Kohannim, O., et al.: Predicting white matter integrity from multiple common genetic variants. Neuropsychopharmacology **37**(9), 2012–2019 (2012)
16. Lee, S.H., et al.: Estimation of pleiotropy between complex diseases using single-nucleotide polymorphism-derived genomic relationships and restricted maximum likelihood. Bioinformatics **28**(19), 2540–2542 (2012)
17. Müllner, D.: Modern hierarchical, agglomerative clustering algorithms. arXiv [stat.ML], September 2011
18. Pichet Binette, A., et al.: Bundle-specific associations between white matter microstructure and ab and tau pathology in preclinical alzheimer's disease. Elife **10**, May 2021
19. Sakoe, H., et al.: Dynamic programming algorithm optimization for spoken word recognition. IEEE Trans. Acoust. **26**(1), 43–49 (1978)
20. Schilling, K.G., et al.: Fiber tractography bundle segmentation depends on scanner effects, vendor effects, acquisition resolution, diffusion sampling scheme, diffusion sensitization, and bundle segmentation workflow. Neuroimage **242**, 118451 (2021)
21. Schroeder, W., et al.: The Visualization Toolkit: An Object-oriented Approach to 3D Graphics. Kitware (2006)
22. Sha, Z., et al.: Genetic architecture of the white matter connectome of the human brain. Sci. Adv. **9**(7), eadd2870 (2023)
23. Smith, S.M., et al.: Tract-based spatial statistics: voxelwise analysis of multi-subject diffusion data. Neuroimage **31**(4), 1487–1505 (2006)
24. Sullivan, C., et al.: PyVista: 3D plotting and mesh analysis through a streamlined interface for the visualization toolkit (VTK). J. Open Source Softw. **4**(37), 1450 (2019)
25. Vuoksimaa, E., et al.: Heritability of white matter microstructure in late middle age: a twin study of tract-based fractional anisotropy and absolute diffusivity indices. Hum. Brain Mapp. **38**(4), 2026–2036 (2017)
26. Yang, J., et al.: Common SNPs explain a large proportion of the heritability for human height. Nat. Genet. **42**(7), 565–569 (2010)
27. Yang, J., et al.: GCTA: a tool for genome-wide complex trait analysis. Am. J. Hum. Genet. **88**(1), 76–82 (2011)
28. Yeh, F.C.: Shape analysis of the human association pathways. Neuroimage **223**, 117329 (2020)
29. Yeh, F.C., et al.: Population-averaged atlas of the macroscale human structural connectome and its network topology. Neuroimage **178**, 57–68 (2018)
30. Yeh, F.C., et al.: Deterministic diffusion fiber tracking improved by quantitative anisotropy. PLoS ONE **8**(11), e80713 (2013)
31. Yeh, F.C., et al.: Generalized q-sampling imaging. IEEE Trans. Med. Imaging **29**(9), 1626–1635 (2010)

32. Yushkevich, P.A.: Continuous medial representation of brain structures using the biharmonic PDE. Neuroimage **45**(1 Suppl), S99-110 (2009)
33. Yushkevich, P.A., et al.: Continuous medial representation for anatomical structures. IEEE Trans. Med. Imaging **25**(12), 1547–1564 (2006)
34. Zhao, B., et al.: Common genetic variation influencing human white matter microstructure. Science **372**(6548), June 2021

Corpus Callosum Parcellation Methods: What Can Tractography Tell Us About Them?

Caio Santana[1]()⍾, Claudio Román[2](), Simone Appenzeller[3](),
Pamela Guevara[4](), and Leticia Rittner[1]()

[1] School of Electrical and Computer Engineering, Universidade Estadual de Campinas (UNICAMP), Campinas, SP, Brazil
c218653@dac.unicamp.br
[2] Centro de I&D en Ingeniería en Salud, Universidad de Valparaíso, Valparaíso, Chile
[3] School of Medical Science, Universidade Estadual de Campinas (UNICAMP), Campinas, SP, Brazil
[4] Universidad de Concepción, Faculty of Engineering, Concepción, Chile

Abstract. The corpus callosum (CC) plays an important role in inter-hemispheric cerebral communication, facilitating the integration and coordination of many brain signals. Its division into subregions, called parcellation, is crucial for studying many brain conditions. However, prevalent parcellation methods rely on fixed geometrical partitioning, failing to adapt to the known variability of the CC across individuals. Data-driven methods based on diffusion MRI have been proposed, but they lack adequate validation. In this study, we used tractography-based analysis to investigate the consistency and compare four CC parcellation methods: Witelson's and Hofer's geometrical approaches, and Cover's and Santana's diffusion MRI-based data-driven methods. Whole-brain tractograms of one hundred subjects from the Human Connectome Project were segmented using the parcellation masks from each method. We then calculated the cortical surface coverage of each segmented tractogram and their inter-subject density correlation coefficients. The average number of streamlines passing through different cortical areas and their relationship with each CC subregion was also assessed. The results revealed significant differences between the methods. Geometrical approaches showed inconsistencies with the expected cortical connections in their subregions. Santana's method presented higher consistency in density correlations and cortical connections, particularly in CC Regions III and IV, which are primarily connected to motor and somatosensory cortical areas. The findings suggest that diffusion MRI-based methods, especially those incorporating directional information, may produce more consistent CC parcellations. Nonetheless, all methods exhibited inconsistencies, resulting in CC subregions that are connected to multiple cortical areas simultaneously.

Keywords: Corpus callosum parcellation · Diffusion Magnetic Resonance Imaging · Tractography

1 Introduction

The corpus callosum (CC) is the largest white matter structure in the human brain and the primary means of interhemispheric cerebral communication. Consisting of a bundle of tightly packed neural fibers located in the middle of the brain, the CC extends from the midline to the cerebral hemispheres through the so-called radiations of the CC. The callosal radiations connect almost all regions of both hemispheres directly or indirectly, facilitating complex integration and coordination of cognitive, sensory, and motor signals [21,29].

Since it plays a significant role in interhemispheric communication, the CC is studied in a variety of conditions, ranging from obsessive-compulsive disorder [16] to schizophrenia [3], multiple sclerosis [5], and Alzheimer's disease [24]. Age and sex differences in the structure have also been found in several studies [9,25].

The axons of the CC exhibit a topographic organization (Fig. 1), with different subregions of the structure presenting different fiber characteristics (i.e., microstructure), such as axonal diameter and degree of myelination [2]. Subregions of the CC also present a functional specialization, according to the cortical brain regions they connect [8].

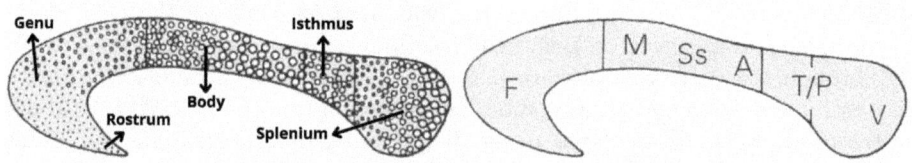

Fig. 1. Representation of anatomical parts (arrows) and fiber diameters along a cross-section of the CC (left). Subregions of the CC associated with different cortical areas (right), from anterior to posterior: F, frontal; M, motor; Ss, somatosensory; A, auditory; T/P, temporoparietal; V, visual. Adapted from [1].

Due to its size and known organization, it is common to subdivide (parcellate) the CC into smaller subregions. This process is generally done in the midsagittal section of the brain (midline) and facilitates the analysis of specific portions of the structure, being crucial for CC-related studies of several neurological diseases and neuropsychiatric disorders. Group differences are often observed only in specific CC subregions and not across the whole structure [16,24,25].

Many methods were proposed to parcellate the CC into subregions, from geometrical [14,31] to data-driven approaches [6,11,20]. They result in distinct CC parcellations, varying in the number of subregions and in their position and size. However, the absence of a gold standard makes it difficult to define the most appropriate method. The geometrical approaches are most widely used, especially Witelson [31] and Hofer [14] partitioning schemes. They subdivide the CC into five regions using fixed proportions of its extent. Therefore, the individualities of each subject's brain are not taken into account. Methods based on imaging data may reduce this problem by using subject-specific information, but further investigations are needed to confirm their applicability and consistency.

Considering the configuration of the CC fibers, it would be expected from parcellation results to be consistent across subjects, identifying parts of the structure that interconnect similar brain regions. Therefore, in this study, we used tractography-based analyses to compare four different CC parcellation methods and assess their consistency between healthy subjects. Beyond Witelson and Hofer geometrical parcellations, we also investigated the Diffusion Tensor Imaging (DTI)-based methods proposed by Cover [6] and Santana [20], since they all divide the CC into five subregions.

2 Materials and Methods

2.1 Data Acquisition and Preprocessing

The dataset used consists of T1w and diffusion MRI acquisitions of 100 subjects from the Human Connectome Project (HCP) [30]. Diffusion data was acquired using b-values of 1000, 2000, and 3000 s/mm^2, with an isotropic voxel size of 1.25 mm. We worked with the HCP pre-processed data [13], which provides diffusion data aligned with the native structural space and a non-linear transformation to MNI152 space. The Dipy library [12] was used to compute the DTI model (eigenvalues and eigenvectors), and the fractional anisotropy (FA) map.

Through the MRtrix3 software [28], fiber orientation distributions (FODs) were estimated using constrained spherical deconvolution (CSD) [27]. Probabilistic tractography was calculated using the second-order Integration over FODs (iFOD2) [26] and the Anatomically-Constrained Tractography (ACT) algorithm [22], generating 30 million streamlines for each subject. Then, Spherical-deconvolution Informed Filtering of Tractograms (SIFT) [23] was applied, resulting in 3 million streamlines per subject. More details of the tractography computation can be found in [19].

2.2 CC Segmentation and Parcellation

The CC volumetric segmentation was performed by the method proposed by Rodrigues et al. [18], based on a U-Net trained on DTI maps. Since the model was trained on a different dataset, we adjusted its threshold and applied morphological closing before extracting the connected components, ensuring the correct segmentation of the structure. Disregarding sagittal slices in the extremities of the brain, the slice with the lowest average FA is chosen as the midsagittal one [10] and defines the midsagittal section of the CC used in the parcellations. Four distinct CC parcellations were obtained for each subject (Fig. 2):

Witelson Parcellation: Witelson's geometric parcellation [31] was one of the first proposed and is still largely used. Based on postmortem connectivity analysis, it subdivides the CC into five vertical sections using fixed proportions of its anterior-posterior line. According to his work, Region I connects

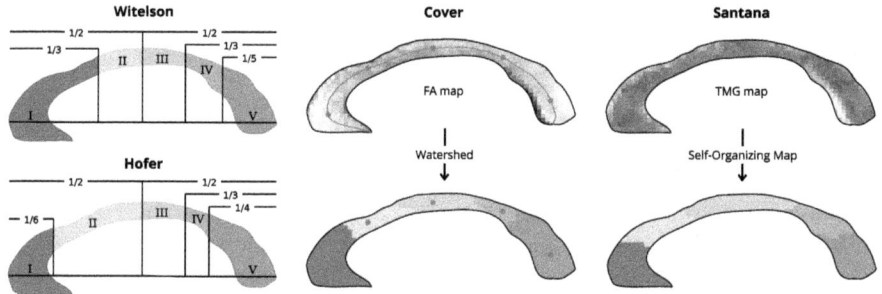

Fig. 2. Parcellation methods: Witelson's and Hofer's geometric schemes (left); Cover's data-driven method based on FA (center); and Santana's data-driven method based on tensorial similarities (right).

prefrontal, premotor, and supplementary motor cortical areas. Region II connects the motor cortex. Region III connects the somaesthetic and posterior parietal areas. Region IV connects the superior temporal and posterior parietal areas. Region V connects the occipital and inferior temporal areas.

Hofer Parcellation: Using DTI-based tractography, Hofer [14] identified subregions of the CC crossed by fibers from specific cortical areas. Considering the average of the studied population, a new geometric partitioning of the structure was defined along its anterior-posterior line. Region I contains fibers from the prefrontal cortex. Region II connects the premotor and supplementary motor cortical areas. Region III comprises fibers from the primary motor cortex. Region IV contains primary sensory fibers. Region V connects the parietal, temporal, and occipital cortices.

Cover Parcellation: Cover [6] proposed an automated data-driven CC parcellation method based on DTI. It applies a k-means algorithm to cluster the points of the central line of the CC based on FA values. The centroids of the five clusters are then used as markers for the Watershed transform, which divides the structure into five subregions. The rationale behind the method is to group voxels with similar diffusion properties.

Santana Parcellation: Santana [20] also proposed an automated data-driven CC parcellation method based on DTI. It uses the Tensorial Morphological Gradient (TMG) map [17] calculated with the Log-Euclidean distance and a 6-connected three-dimensional structuring element. Data from the two neighbor slices is also used to calculate the TMG of the midsagittal section of the CC, considering all the tensorial information available instead of discarding the directional information as in the FA. The TMG values and their spatial coordinates are used as input to a Self-Organizing Map, which clusters the data to define five CC subregions. Since the TMG highlights tensorial differences, the method looks for regions containing similar diffusion tensors.

Parcellation methods from Witelson, Hofer, and Cover were computed using the inCCsight tool [4]. Santana's parcellation method [20] was computed using the Python code available on GitHub.[1]

2.3 Tractography-Based Analysis

The parcellation masks were used to segment the whole-brain tractograms, resulting in individual tractograms for each CC subregion of each parcellation method per subject. Subsequently, they were transformed to MNI space.

Density Maps and Density Correlation: The tractograms were converted to density images, in which the value of each voxel represents the number of fibers passing through it. To assess the similarity between the density images, we calculated the correlation between the density images of each pair of subjects. Then, we averaged the correlation coefficients to obtain mean density correlations for each subregion of each parcellation. The density correlation is defined as the Pearson correlation between a pair of bundles (X and Y):

$$\rho_{X,Y} = \frac{\text{cov}(X,Y)}{\sigma_X \sigma_Y} \quad (1)$$

Surface Coverage: To compare the cortical surface coverage of the different parcellations, we obtained the intersection between the triangles of the cortical surface mesh and the tractograms of each CC subregion. Then, the percentage of intersected triangles was calculated.

To determine the cortical regions related to each CC subregion, we calculated the average number of streamlines ending in each area of the Desikan-Killiany cortical atlas [7]. They were grouped into frontal, motor, somatosensory, posterior parietal, temporal, occipital, and insular cortices. Then, we obtained the percentage of streamlines of each cortical area passing through each CC subregion.

3 Results

As expected, tractography shows that streamlines from each CC subregion tend to be concentrated in specific cortical areas (Fig. 3). However, these areas vary across subjects and some overlap and scattering of streamlines are also observed.

The cortical surface coverage of each CC subregion differs between the methods (Fig. 4). In general, higher coverage and inter-subject variation are found in Region V, although Santana's parcellation presents smaller percentages than others. Witelson's method shows a particularly higher coverage in Region I and Hofer's in Region II. The coverage of Regions III and IV tends to be higher for Santana's method, which also exhibits more similar values between its subregions. Cover's method tends to present more variability in its results.

All studied parcellation methods exhibit low to moderate density correlations (Fig. 4), with averages ranging between 0.2 and 0.6. In Region I and Region II,

---

[1] https://github.com/MICLab-Unicamp/TMG-based_CC_Parcellation.

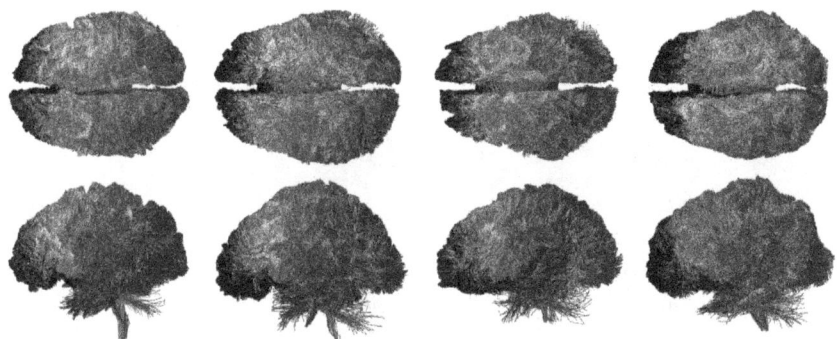

Fig. 3. Tractography results of four subjects (columns) using Santana's parcellation method. Axial views (up) and sagittal views (down).

Santana's and Hofer's parcellations present higher correlation coefficients. For Regions III and IV, Santana's parcellation has the highest correlation coefficients, while the lowest correlations are found in Hofer's method. In Region V, Witelson's and Hofer's parcellations present higher density correlations.

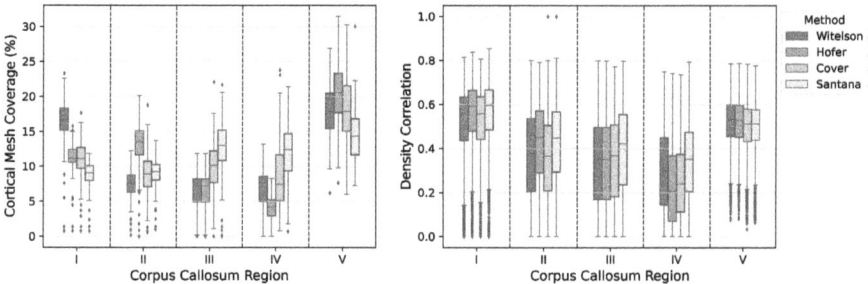

Fig. 4. Percentages of cortical surface coverage (left) and pairwise density correlation coefficients (right) per CC subregion and parcellation method.

Santana's parcellation presents the highest overall mean correlation (0.441 ± 0.187), followed by Hofer (0.411 ± 0.207), Witelson (0.409 ± 0.196), and Cover (0.395 ± 0.195). In general, Regions V and I show higher density correlations and lower variation, while Region IV exhibits lower correlations and Regions II and III present higher variation.

The relationship between CC subregions and cortical regions tends to follow a similar overall pattern throughout all the parcellation methods (Fig. 5). Still, there are interesting variations worth noting.

Region I of Witelson's method comprises most of the streamlines passing through the frontal cortex, while Santana's method shows a lower proportion of its streamlines. Cover's and Santana's parcellations exhibit similar results

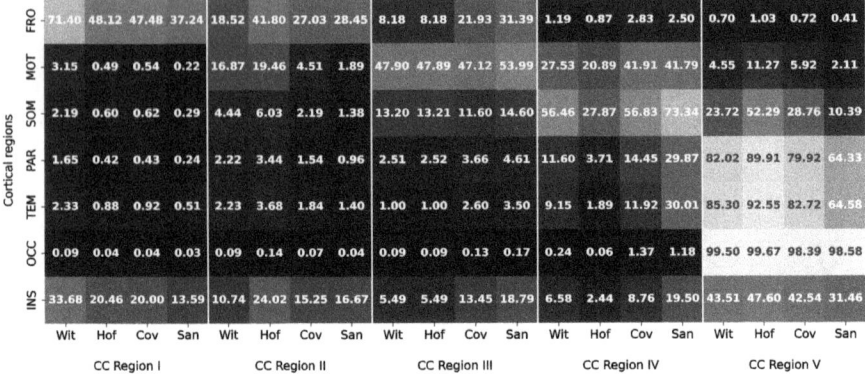

Fig. 5. Percentage of streamlines in each cortical region (rows) that pass through each CC subregion (columns). FRO, frontal; MOT, motor; SOM, somatosensory; PAR, posterior parietal; TEM, temporal; OCC, occipital; INS, insula; Wit, Witelson; Hof, Hofer; Cov, Cover; San, Santana.

in Region II, dominated by frontal streamlines, while Hofer's and Witelson's methods include motor streamlines as well.

Region III presents a higher proportion of motor cortex and some somatosensory streamlines, especially in Santana's method. Cover's and Santana's methods show a higher percentage of frontal streamlines in this subregion. Region IV of Hofer's parcellation contains streamlines from the motor and somatosensory areas. This holds for the other methods, but they also tend to present a proportion of posterior parietal and temporal streamlines, especially in Santana's method.

Almost all occipital streamlines pass through CC Region V, which presents most of the posterior parietal and temporal streamlines. Most somatosensory and some motor streamlines are found in this subregion on Hofer's parcellation. Streamlines of the insular cortex tend to concentrate in this subregion but are also highly present in Regions I and II. Santana's parcellation shows insular streamlines passing through all its subregions.

4 Discussion

Overall Differences. Results indicate a clear difference between parcellation methods. While this can be seen in their definition (Fig. 2), the cortical coverage (Fig. 4) and connectivity analyses (Fig. 5) highlight this even more, since streamlines filtered using different CC parcellation techniques demonstrate different patterns of cortical connectivity. These differences are related to the size of each CC subregion. For example, Region I of Witelson's method tends to be larger compared to the same subregion in other methods, resulting in higher coverage and a higher proportion of frontal streamlines passing through it. The same goes for Regions II and V of Hofer's parcellation and Regions III and IV of Santana's method, although for different cortical regions. The opposite is also verified, with the smaller Regions I and V of Santana's method showing lower coverage and percentage of streamlines, which also happens in Region IV of Hofer's method.

Density Correlation. The density correlation analysis indicates a higher intersubject similarity with lower variation for the tractograms segmented by Santana's parcellation. In particular, this method presented the highest correlation coefficients for Regions III and IV, which tended to be the subregions with smaller correlations throughout the methods. On the other hand, Cover parcellation presented the lowest overall density correlation. It should be noted, however, that the differences found between the methods tended to be small.

Moreover, all parcellation methods exhibit low to moderate density correlations. This may be related to the variability between each subject's brain, not only in the CC itself but also in the cortical surface and the organization of the neural fibers. As indicated by Huang et al. [15], the tracts that traverse the CC are highly variable across subjects and only its most anterior and posterior areas present a high inter-subject consistency. We found Regions I and V to present higher correlation coefficients in all parcellation methods, which is consistent with Huang's results. The tractography algorithm may also contribute to the low correlation results since false positive streamlines can artificially increase inter-subject variability. Nonetheless, the analysis used the same set of 100 tractograms registered to a common space, with variations only in their segmentation. Therefore, differences between the methods are not attributable to the tractography algorithm itself but rather indicate variations in inter-subject consistency among the different parcellation methods.

Geometrical Methods' Connectivity. The analysis of the percentage of streamlines in each cortical area that passes through each CC subregion shows divergences concerning what would be expected for the geometrical methods. According to Witelson's work [31], CC Region II should be connected to the motor cortex, but we observed a small proportion of streamlines from this area and some streamlines from the frontal cortex. Region III was the one to present the highest percentage of motor streamlines, even though it should be focused on somatosensory and posterior parietal areas. Somatosensory streamlines were also found in Region III but in a smaller proportion compared to Region IV, which was expected to be connected to superior temporal and posterior parietal cortical areas. Region IV also showed a considerable amount of motor streamlines. Although presenting most of the streamlines from occipital and temporal areas as expected, Region V of Witelson's method also contains most of the posterior parietal streamlines and even some from the somatosensory cortex.

Regarding Hofer's parcellation [14], Region I should be connected to the prefrontal cortex and Region II to the premotor and supplementary motor areas. Although we cannot confirm this directly, it seems to make sense, since streamlines from the frontal cortex are divided between these two subregions. However, Region II also presents a percentage of streamlines from the primary motor area. Region III should be connected to the primary motor cortex and it indeed presents the most concentration of streamlines from this cortical area, although somatosensory streamlines are also found in this subregion. Moreover, primary motor streamlines can also be found in Region IV together with somatosensory ones, but this subregion was supposed to be connected only to the somatosensory cortex. Finally, Region V should be connected to the parietal, temporal, and occipital cortices, but most of the streamlines from the somatosensory cortex were found in this subregion, which also presents connections with the motor area.

Data-Driven Methods. For the data-driven methods, there is no specific indication of which cortical areas should be connected to which CC subregion. We observed that Regions I and II of both Cover's and Santana's parcellations present streamlines only from the frontal cortex, with Cover's showing a higher percentage of frontal streamlines within these two subregions. In contrast to the geometrical methods, a higher proportion of frontal connections can also be found in Region III. Santana's method shows the highest proportion of motor and somatosensory streamlines in a single Region (III and IV, respectively). However, these subregions are also connected to other cortical areas, with a high percentage of streamlines from the frontal cortex in Region III and from the motor, parietal, and temporal cortices in Region IV. Moreover, this method shows the lowest concentration of parietal and temporal connections in Region V.

The reasoning behind Cover's method is to identify CC subregions with distinct microstructural characteristics using the FA map. While FA can approximate some microstructural features, it may not ensure a coherent CC parcellation, particularly because it discards all directional tensorial information. However, only a histological analysis could confirm if Cover's method finds CC subregions with coherent internal microstructural characteristics. In contrast, Santana's method subdivides the CC using the complete tensorial information from the TMG map, clustering similar tensors together. This method considers the directionality of diffusion tensors, seeking fibers that traverse the midsagittal section of the CC in similar directions. Consequently, Santana's parcellation may have an advantage in identifying CC subregions that are coherent both microstructurally and in terms of cortical connections. This is partially supported by our results, as Santana's parcellation demonstrated higher density correlations and CC subregions with a higher proportion of specific cortical connections, especially in Regions III and IV. Nevertheless, Santana's method still showed considerable variation between subjects (Fig. 3).

5 Conclusion

Although the CC has a topographical organization, it exhibits significant variability across subjects and changes throughout the lifespan. Thus, fixed geometrical partitioning of the structure is likely to result in inconsistent CC subregions.

Our results seem to support this hypothesis, as the relationship between the CC subregions and the cortical areas differed from expectations for Hofer's and Witelson's geometrical parcellation methods.

Diffusion MRI has the potential to reveal microstructural white matter characteristics and cerebral connectivity, potentially leading to more consistent CC parcellation methods that can adapt to individual brain features. Our analyses suggest that Santana's approach yields a slightly more consistent parcellation, whereas Cover's method does not show the same level of consistency. This indicates that the directional information provided by DTI is likely an important factor in achieving a coherent CC subdivision.

The reasoning for parcellating the CC comes from its known organization and the functional specialization of its subregions. Therefore, analyses of the CC subregions can reveal specific alterations that may not be evident in the entire structure, allowing researchers to associate these findings with particular brain pathways and processes. To achieve this, parcellation methods should ideally define CC subregions consistently connected to specific cortical areas across subjects. Instead, we found all parcellation methods to present inconsistencies and generate CC subregions connected to multiple cortical areas simultaneously. This, along with the lack of a universally adopted parcellation method, can obscure true group differences or create artificial ones, making comparisons across studies difficult and contributing to inconsistent findings in the literature. Furthermore, variability in the connectivity of CC subregions makes it challenging to draw clear conclusions and associate CC alterations with brain functioning.

It should be noted, however, that our analyses were based on a single dataset with high spatial and angular resolution. Therefore, further investigations are necessary to confirm our findings. The choice of the tractography reconstruction algorithm and its configuration may also affect the analysis. It would be interesting to assess their influence on the results by comparing different options. Additionally, variations in the segmentation of the CC could impact its subdivision, especially for Cover's and Santana's methods. Finally, the cortical parcellation may also influence the results, and a finer cortical subdivision could lead to a better assessment of the connectivity of CC subregions.

Acknowledgments. This study was financed in part by the Coordenação de Aperfeiçoamento de Pessoal de Nivel Superior - Brasil (CAPES) - Finance Code 001. Caio Santana and Leticia Rittner thank the National Council of Scientific and Technological Development (CNPq grants #141079/2021-5 and #317133/2023-3). Pamela Guevara and Claudio Román thank the National Research and Development Agency ANID (Basal Project FB0008 (AC3E) to PG and CR, and FONDECYT Postdoctorado 3220729 to CR).

Disclosure of Interests. The authors have no competing interests to declare.

References

1. Aboitiz, F., Montiel, J.: One hundred million years of interhemispheric communication: the history of the corpus callosum. Braz. J. Med. Biol. Res. **36**, 409–420 (2003). https://doi.org/10.1590/S0100-879X2003000400002
2. Aboitiz, F., Scheibel, A.B., Fisher, R.S., Zaidel, E.: Fiber composition of the human corpus callosum. Brain Res. **598**(1), 143–153 (1992). https://doi.org/10.1016/0006-8993(92)90178-C
3. Arnone, D., McIntosh, A.M., Tan, G., Ebmeier, K.P.: Meta-analysis of magnetic resonance imaging studies of the corpus callosum in schizophrenia. Schizophr. Res. **101**(1), 124–132 (2008). https://doi.org/10.1016/j.schres.2008.01.005
4. Caldeira, T., Julio, P.R., Appenzeller, S., Rittner, L.: inCCsight: a software for exploration and visualization of DT-MRI data of the Corpus Callosum. Comput. Graph. **99**, 259–271 (2021). https://doi.org/10.1016/j.cag.2021.07.012
5. Caverzasi, E., Papinutto, N., Cordano, C., Kirkish, G., Gundel, T.J., Zhu, A., et al.: MWF of the corpus callosum is a robust measure of remyelination: Results from the ReBUILD trial. PNAS **120**(20), e2217635120 (2023). https://doi.org/10.1073/pnas.2217635120
6. Cover, G., Pereira, M., Bento, M., Appenzeller, S., Rittner, L.: Data-driven corpus callosum parcellation method through diffusion tensor imaging. IEEE Access **5**, 22421–22432 (2017). https://doi.org/10.1109/ACCESS.2017.2761701
7. Desikan, R.S., Ségonne, F., Fischl, B., Quinn, B.T., Dickerson, B.C., Blacker, D., et al.: An automated labeling system for subdividing the human cerebral cortex on MRI scans into gyral based regions of interest. Neuroimage **31**(3), 968–980 (2006). https://doi.org/10.1016/j.neuroimage.2006.01.021
8. Fabri, M., Polonara, G.: Functional topography of human corpus callosum: an fMRI mapping study. Neural Plast. **2013**(1), 251308 (2013). https://doi.org/10.1155/2013/251308
9. Fan, Q., Tian, Q., Ohringer, N.A., Nummenmaa, A., Witzel, T., Tobyne, S.M., et al.: Age-related alterations in axonal microstructure in the corpus callosum measured by high-gradient diffusion MRI. Neuroimage **191**, 325–336 (2019). https://doi.org/10.1016/j.neuroimage.2019.02.036
10. Freitas, P., Rittner, L., Appenzeller, S., Lapa, A., Lotufo, R.: Watershed-based segmentation of the corpus callosum in diffusion MRI. In: Medical Imaging 2012: Image Processing. vol. 8314, pp. 879–885. SPIE (2012). https://doi.org/10.1117/12.911619
11. Friedrich, P., Fraenz, C., Schlüter, C., Ocklenburg, S., Mädler, B., Güntürkün, O., et al.: The relationship between axon density, myelination, and fractional anisotropy in the human corpus callosum. Cereb. Cortex **30**(4), 2042–2056 (2020). https://doi.org/10.1093/cercor/bhz221
12. Garyfallidis, E., Brett, M., Amirbekian, B., Rokem, A., Van Der Walt, S., Descoteaux, M., et al.: Dipy, a library for the analysis of diffusion MRI data. Front. Neuroinform. **8** (2014). https://doi.org/10.3389/fninf.2014.00008
13. Glasser, M.F., Sotiropoulos, S.N., Wilson, J.A., Coalson, T.S., Fischl, B., Andersson, J.L., et al.: The minimal preprocessing pipelines for the Human Connectome Project. Neuroimage **80**, 105–124 (2013). https://doi.org/10.1016/j.neuroimage.2013.04.127
14. Hofer, S., Frahm, J.: Topography of the human corpus callosum revisited-Comprehensive fiber tractography using diffusion tensor magnetic resonance imaging. Neuroimage **32**(3), 989–994 (2006). https://doi.org/10.1016/j.neuroimage.2006.05.044

15. Huang, H., Zhang, J., Jiang, H., Wakana, S., Poetscher, L., Miller, M.I., et al.: DTI tractography based parcellation of white matter: application to the mid-sagittal morphology of corpus callosum. Neuroimage **26**(1), 195–205 (2005). https://doi.org/10.1016/j.neuroimage.2005.01.019
16. Piras, F., Vecchio, D., Kurth, F., Piras, F., Banaj, N., Ciullo, V., et al.: Corpus callosum morphology in major mental disorders: a magnetic resonance imaging study. Brain Commun. **3**(2), fcab100 (2021). https://doi.org/10.1093/braincomms/fcab100
17. Rittner, L., Campbell, J.S.W., Freitas, P.F., Appenzeller, S., Bruce Pike, G., Lotufo, R.A.: Analysis of Scalar Maps for the Segmentation of the Corpus Callosum in Diffusion Tensor Fields. J Math Imaging Vis **45**(3), 214–226 (2013). https://doi.org/10.1007/s10851-012-0377-4
18. Rodrigues, J., Pinheiro, G., Carmo, D., Rittner, L.: Volumetric segmentation of the corpus callosum: training a deep learning model on diffusion MRI. In: 17th Int. Symp. Med. Inf. Proc. (SIPAIM), vol. 12088, pp. 198–207. SPIE (2021). https://doi.org/10.1117/12.2606233
19. Román, C., Hernández, C., Figueroa, M., Houenou, J., Poupon, C., Mangin, J.F., et al.: Superficial white matter bundle atlas based on hierarchical fiber clustering over probabilistic tractography data. Neuroimage **262**, 119550 (2022). https://doi.org/10.1016/j.neuroimage.2022.119550
20. Santana, C., Abreu, T., Rodrigues, J., Julio, P., Appenzeller, S., Rittner, L.: DTI-based Corpus Callosum parcellation using the Tensorial Morphological Gradient and Self-Organizing Maps. In: 2023 19th Int. Symp. Med. Inf. Proc. (SIPAIM) pp. 1–5 (2023). https://doi.org/10.1109/SIPAIM56729.2023.10373443
21. Shah, A., Jhawar, S., Goel, A., Goel, A.: Corpus callosum and its connections: a fiber dissection study. World Neurosurgery **151**, e1024–e1035 (2021). https://doi.org/10.1016/j.wneu.2021.05.047
22. Smith, R.E., Tournier, J.D., Calamante, F., Connelly, A.: Anatomically-constrained tractography: improved diffusion MRI streamlines tractography through effective use of anatomical information. Neuroimage **62**(3), 1924–1938 (2012). https://doi.org/10.1016/j.neuroimage.2012.06.005
23. Smith, R.E., Tournier, J.D., Calamante, F., Connelly, A.: SIFT: spherical-deconvolution informed filtering of tractograms. Neuroimage **67**, 298–312 (2013). https://doi.org/10.1016/j.neuroimage.2012.11.049
24. Sydykova, D., Stahl, R., Dietrich, O., Ewers, M., Reiser, M.F., Schoenberg, S.O., et al.: Fiber connections between the cerebral cortex and the corpus callosum in Alzheimer's disease: a diffusion tensor imaging and voxel-based morphometry study. Cereb. Cortex **17**(10), 2276–2282 (2007). https://doi.org/10.1093/cercor/bhl136
25. Tanaka-Arakawa, M.M., Matsui, M., Tanaka, C., Uematsu, A., Uda, S., Miura, K., et al.: Developmental changes in the corpus callosum from infancy to early adulthood: a structural magnetic resonance imaging study. PLoS ONE **10**(3), e0118760 (2015). https://doi.org/10.1371/journal.pone.0118760
26. Tournier, J.D., Calamante, F., Connelly, A.: Improved probabilistic streamlines tractography by 2nd order integration over fibre orientation distributions. In: Proc. of the International Society for Magnetic Resonance in Medicine. ISMRM (2010)
27. Tournier, J.D., Calamante, F., Connelly, A.: Robust determination of the fibre orientation distribution in diffusion MRI: Non-negativity constrained super-resolved spherical deconvolution. Neuroimage **35**(4), 1459–1472 (2007). https://doi.org/10.1016/j.neuroimage.2007.02.016

28. Tournier, J.D., Calamante, F., Connelly, A.: MRtrix: diffusion tractography in crossing fiber regions. Int. J. Imaging Syst. Technol. **22**(1), 53–66 (2012). https://doi.org/10.1002/ima.22005
29. Turgut, M., Tubbs, R.S., Turgut, A.T., Bui, C.C. (eds.): The Corpus Callosum: Embryology, Neuroanatomy, Neurophysiology, Neuropathology, and Surgery. Springer (2023). https://doi.org/10.1007/978-3-031-38114-0
30. Van Essen, D., Ugurbil, K., Auerbach, E., Barch, D., Behrens, T., Bucholz, R., et al.: The human connectome project: a data acquisition perspective. Neuroimage **62**(4), 2222–2231 (2012). https://doi.org/10.1016/j.neuroimage.2012.02.018
31. Witelson, S.F.: Hand and sex differences in the isthmus and genu of the human corpus callosum: a postmortem morphological study. Brain **112**(3), 799–835 (1989). https://doi.org/10.1093/brain/112.3.799

Author Index

A
Adluru, Nagesh 164
Aja-Fernández, Santiago 164
Amorosino, Gabriele 95
Appenzeller, Simone 210
Archer, Derek 132
Avesani, Paolo 95

B
Ba Gari, Iyad 197
Bach Cuadra, Meritxell 24
B. McNabb, Carolyn 106
Bærentzen, J. Andreas 35
Bao, Shunxing 132
Barritt, Andrew W. 143
Bhatt, Ravi R. 197
Bouyagoub, Samira 143

C
C. Alexander, Daniel 106
Cai, Leon Y. 132
Cai, Weidong 84
Cercignani, Mara 106
Chamberland, Maxime 12, 72, 164
Chandio, Bramsh Qamar 164
Chen, Geng 1
Chen, Yufei 24
Chen, Yuqian 84
Chen, Zijian 119
Cheng, Jian 60, 153
Cicimen, Alp G. 106
Ciupek, Dominika 164
Consagra, William 164
Cui, Can 132
Cui, Ruiqi 35

D
De Baene, Wouter 12
Deng, Ruining 132
Dong, Enqing 153
Dyrby, Tim B. 35

E
Evans, C. John 106

F
Fan, Wenxin 60, 153
Figini, Matteo 106
Florack, Luc 12
Ford, Alexandra 143

G
Gade, Anurag 164
Garyfallidis, Eleftherios 164
Genc, Sila 164
Ghezzi, Sofia 95
Gholipour, Ali 24
Golby, Alexandra J. 84
Gruen, Johannes 175
Guevara, Pamela 210

H
Hendriks, Tom 164
Huo, Yuankai 132

J
Jahanshad, Neda 164, 197
Jiang, Haotian 1
Jiménez, Gabriela Gómez 185
Jones, Derek K. 106
Jovicich, Jorge 95

K
Kanakaraj, Praitayini 132, 164
Karimi, Davood 24
Kebiri, Hamza 24
Kelly, Claire E. 164
Koudoro, Serge 164

L

Landman, Bennett A. 132
Landman, Bennett 164
Li, Cheng 60, 153
Li, Zhiyuan 132
Lin, Rizhong 24
Liu, Dongnan 84
Liu, Feihong 1
Liu, Quan 132
Liu, Wan 84
Lo, Yui 84

M

Ma, Jiquan 1
Machnio, Julia 164
Makris, Nikos 84
Malawski, Maciej 164
Messaritaki, Eirini 106
Moreno, Rodrigo 47
Moyer, Daniel 164

N

Nath, Vishwesh 164
Newlin, Nancy 164, 132

O

O'Donnell, Lauren J. 84
Ouedraogo, Gani 164

P

Palombo, Marco 106
Pathak, Sudhir 164
Persson, Sanna 47
Pieciak, Tomasz 164

R

Rathi, Yogesh 84, 164
Rekik, Islem 1
Riccardi, Chiara 95
Rittner, Leticia 210
Román, Claudio 210
Rutten, Geert-Jan 12

S

Santana, Caio 210
Sarubbo, Silvio 95
Schilling, Kurt 132, 164
Schneider, Walter 164
Schultz, Thomas 175
Shen, Dinggang 1
Smolders, Lars 12

T

Teng, Yujun 1
Thiran, Jean-Philippe 24
Thompson, Paul M. 164
Tregidgo, Henry F. J. 106

V

van der Hofstad, Remco 12
Vega, Antonio Tristán 164
Venkataraman, Archana 119
Vilanova, Anna 72, 164
Vink, Ruben 72

W

Wan, Xinyi 47
Wang, Jueqi 119
Wang, Shanshan 60, 153
Wassermann, Demian 185
Wu, Ruoyou 60
Wu, Ye 164

X

Xiao, Taohui 153

Y

Yang, Jing 60, 153
Yang, Joseph Yuan-Mou 164
Yao, Tianyuan 132
Yeh, Fang-Chang 197

Z

Zekelman, Leo 84
Zhang, Fan 84
Zhang, Kai 1
Zigiotto, Luca 95
Zou, Juan 60

GPSR Compliance

The European Union's (EU) General Product Safety Regulation (GPSR) is a set of rules that requires consumer products to be safe and our obligations to ensure this.

If you have any concerns about our products, you can contact us on

ProductSafety@springernature.com

In case Publisher is established outside the EU, the EU authorized representative is:

Springer Nature Customer Service Center GmbH
Europaplatz 3
69115 Heidelberg, Germany

www.ingramcontent.com/pod-product-compliance
Lightning Source LLC
Chambersburg PA
CBHW071942050525
26154CB00014B/130